TAKEOVER

A VEX Robotics Novel

By Nathan Miles

BRAINSTEM
BOOKS

Contributions by
VRC Team 10703Z BrainSTEM Botz

VEX is the registered trademark of VEX Robotics, Inc. and Innovation First, Inc.

Foreword

Don't you just love a good story filled with twists and turns? And with plots that leave you on the edge of your seat in suspense, wondering what will happen next? The action and drama make your heart race. You share moments of elation, devastation, confusion, and determination, as well as a full range of emotions with each character. A well-crafted story can be intense and exciting, and so much fun that you are left longing for more. What if this level of fun and excitement could be *real*? What if you could step into the action and become an actual part of the story?

Welcome to VEX Robotics.

It starts with a drive and determination to build the most competitive robot possible. It involves countless hours of work, knowing the rules of the game, keeping an engineering notebook, and working with screws, gears, programming, people, strategy... the list is endless. Efforts made in robotics come with emotional highs and lows produced by successes and failures. The path of robotics is adventurous, much like stepping right into a story in which ***you*** are a leading character.

If you have been involved with VEX Robotics, you know how much fun tournaments can be. You know how satisfying it can be to see your work come to fruition after an insane number of hours spent working. You know what it's like to be part of a team where members have strong personalities and differing ideas. You have learned how to build a robot and may have even gained experience with CAD (computer aided design), sensors, and programming. As you read this book, I hope you will enjoy reliving your adventures in the world of VEX Robotics.

For those unfamiliar with VEX Robotics, a new game is announced annually. Tower Takeover was the game for 2019-2020 and was played on a 12' x 12' playing field (pictured on the front cover of this book). The field elements consisted of a scoring zone in each corner and seven towers of various heights in designated positions on the field. Sixty-six cubes, each one green, orange, or purple, were arranged around the field. Four teams, two on the red alliance and two on the blue alliance, would play simultaneously. Each match would start with an autonomous period of fifteen seconds (where the robot's movements were pre-programmed), followed by one minute and forty-five seconds of driver control. To score, each team would have their robot stack cubes in the scoring zones. Cubes placed in towers would multiply the point value for that color of cube in the scoring zones.

Students typically become involved in VEX Robotics by joining an existing team through a school, homeschool group, or community organization. A coach or other mentor officially registers each team with REC Foundation at robotevents.com. The coach plays an important role by signing up the team for various competitions and making sure the students have parts and tools as needed. Beyond these types of administrative responsibilities, REC Foundation requires all teams to be student-centered. Coaches may encourage and teach, but students must do all of the work themselves. Teams meet regularly to build their robot and document their progress in an engineering notebook. This, along with practice and programming, prepares each team for tournaments. If you are considering giving competitive robotics a try, I have two words for you: Do. It. No matter what grade you are in, find a team you can join, or try to start a new team! VEX Robotics competitions are the experience of a lifetime and something you will always treasure. And, you will learn valuable lessons in STEM, problem-solving, and interpersonal skills along the way which will benefit you for the rest of your life.

After being involved in VEX Robotics for years as a supportive parent and coach, I have seen firsthand just how beneficial and educational competitive robotics can be for students. It is truly remarkable how much students can learn when STEM is turned into a sport and the educational process takes the form of healthy competition.

Thank you to VEX Robotics for allowing the use of their name and logo in this book. With the exception of the VEX name and logo, VEX has not approved the content of this book. Also, thank you to REC Foundation (Robotics Education Competition) for their support and encouragement as this book was written. REC Foundation works closely with VEX Robotics to provide robotics opportunities to students in all grades, along with managing an endless list of responsibilities that accompany this immense endeavor.

Most of all, thank you to my wonderful family. Thank you to my wife who spent countless hours editing the text of this book. And thank you to my daughter, from VRC Team 10703Z BrainSTEM Botz, who assisted with the storyline and helped make sure that the setting of VEX Robotics in the book is as realistic as it can be. Without my family jumping in and playing these crucial roles, this book would not have been possible.

This story is about a fictional team competing during the Tower Takeover season, and the purpose of this book is to give you a taste of the adventure of VEX Robotics, the hard work it encompasses, the excitement of giving your best, and the wonderful rewards that can come from learning and competing. If you like *Takeover*, watch for a sequel in 2021 based on Change Up, the VEX Robotics 2020-2021 challenge. Information about new titles from BrainSTEM Books may be found at:

brainstembooks.info/books

The following robotics story is fictional. Any resemblance in this story to an actual person, team, tournament, etc. is unintentional and coincidental. However, if you have ever been involved in VEX Robotics, you may occasionally smile with the realization that, *Yeah, I've met someone just like that*. Or, *Yep, that has happened to my team.*

With this introduction, let the story begin as we step into the grand adventure of VEX Robotics.

Nathan Miles

Chapter 1

It was a moment like no other! So much had happened in the last few hours, it was almost impossible to believe. Kenzie glanced up at the big screen to confirm that this was *not* a dream. Yep. The filled-in brackets made it official. The Robonauts from little ole Valley Creek High School had actually gone all the way to the final elimination match at their robotics tournament, Tumultuous Towers. The next few moments would determine the Tournament Champion.

"Now remember," said Kenzie, looking over at their allied team, "you two focus on putting purple cubes in the low towers, and we'll do our best to stack purple in our scoring zones. OK?"

Luis stared back at her from across the queuing table with a firm nod of the head. He was obviously nervous. Kenzie just hoped he could follow through on this basic strategy while under pressure. Luis and Loren were identical twins and next to impossible to tell apart. Both were thin, had dark complexions, and wore the same team shirt with their fighter jet logo on it. The one identifying feature was Luis' wide-rimmed glasses. Together these twins comprised the Sonic Boomers. The Sonic Boomers were a new team and still learning everything a team needs to know about robotics tournaments, competing, and strategy in general. They definitely needed help with their team name. But despite their inexperience, this team had still managed to rank fifth in the qualification matches.

The Sonic Boomers' tiny clawbot was ironically named Goliath. Goliath had a claw that could pick up one cube at a time and raise it up into a tower. But the strongest quality of Goliath was its ability to play defense. It could ram hard into a robot - which was not very welcome to a robot in the process of balancing a stack of cubes. Luis also had a knack for positioning Goliath right in the way of opposing robots. Furthermore, whenever the robot stopped, its wheels would lock, preventing the robot from getting pushed out of the way very easily.

Kenzie and her team, the Robonauts, had named their robot Columbia after the first space shuttle NASA had sent into space that was able to orbit the earth, return, land, and be redeployed again on future missions way back in the 1980s.

Kenzie, along with Emma and Corbin, her teammates, lined the other side of the queuing table with Columbia sitting on the table in the middle. Without leaving their designated spot, all three were standing on their tiptoes, stooping, or doing whatever they could to see through the crowd surrounding the field in the center of the gym. There, the last semifinal tournament match was in progress and almost over.

"Has anyone seen Coach Warren?" asked Kenzie, slowly scanning back and forth over the bleachers.

Coach Warren was also the team's official videographer. He had been moving around the gym all day long, taking videos of each match in which the Robonauts competed. His videos would be studied later by the team, especially the video of their final match, which would begin shortly.

Emma was the first to spot him standing over at the side of the gym. "There he is, talking with Curly."

"Now what conversation with Curly could be more important than taking a video of our match?" remarked Kenzie.

"Who knows," replied Emma. "He's been talking with Curly off and on all day."

It's amazing how you tend to see the same people over and over at tournaments and start to feel like you know them, even though you only interact with them briefly or perhaps just see them in passing. Most likely such people you encounter are other students, coaches, parents, volunteers, or possibly a judge coming around to do a team interview. But all have a common love of robotics, which makes for a bit of an unspoken, but special bond. In lieu of not always knowing people's names, Kenzie and Emma would sometimes come up with nicknames. Hence the name, Curly, since this guy, whoever he was, had long black hair with some of the biggest curls these girls had ever seen, always topped off with a red baseball cap.

"There he goes," said Emma, as she watched their coach change his focus and head over to take a seat on the top row of the bleachers, right next to the field where they would soon be competing.

Looking down from the bleachers, Coach Warren gave a thumbs up sign to the Robonauts, indicating to them that he had not forgotten and would capture the ever important video of this final match.

It had been a frustrating day because, for some reason, every so often in the middle of a match, something would go wrong with a robot on the field. It would simply freeze and become unresponsive. The robot would start working again only after the match and once the team could restart it. There seemed to be no explanation as to what was causing this, when it happened, or who it affected. One coach had suggested that maybe the problem was with static electricity buildup on the field which can do bad things to a robot. One of the scorekeepers speculated that something was wrong with Tournament Manager, the software used to enable/disable the robots on the field. But no one knew for sure what was going on. This problem might not be so bad if it affected all robots on the field equally or if there was time for these matches to be replayed. But it seemed that only one robot here or there would encounter this problem. And each of these robots was frozen at the worst possible time, right as a stack of cubes was being delivered or right as a cube was about to be placed in a tower. It was aggravating for everyone involved. Indeed, Columbia had frozen once earlier in the day right as Emma, the Robonauts' driver, was delivering a stack of cubes. Corbin, the Robonauts' programmer, had been pulling his hair out all day long trying to figure out what was going on and if anything in his code had contributed to Columbia's freezing.

A loud buzzer went off, indicating the end of the last semifinal match. The scorekeepers counted the points - twice, just to be certain. Then, after the final scores had been confirmed with each participating team, the MC picked up his microphone.

"And the winner is..."

The MC always paused dramatically before announcing the winning alliance. "Teams Heavy Metal and The Best." The crowd in the stands cheered wildly and a small section of supporters from team Heavy Metal broke out screaming and stomping.

The team named The Best was extremely annoying. Primarily because they truly thought they were the best. And they loved to say so. The girl who drove for The Best always made a point to introduce her team to everyone by saying, "Hey y'all, we're *The Best.*" The frustrating part was that this team was actually very good. The Best had designed their robot so that it stacked cubes in a unique way, from the bottom up. Basically, the robot would start by delivering the first cube to the ground. Then it had a DR4B lift that would grab this cube, raise it up in the air, use intakes running in reverse to push a second cube underneath, and then lower the first cube back down on top of it. After this, the robot would then raise up the two stacked cubes and push yet a third cube under the stack. Then a fourth. Then a fifth. It was crazy! The most Kenzie had seen them stack all day was nine cubes.

Team Heavy Metal's robot wasn't bad either. Their robot, a simple traybot that could stack maybe four to five cubes at the most, was fairly solid and consistent. Their biggest weakness, though, seemed to be sloppy driving. The driver often seemed to be in such a rush to complete whatever task was at hand that the robot's movements were sometimes jerky and erratic.

Waiting for the pandemonium to subside a bit, the MC continued, "Would our two alliances competing in the final match please make your way to the field?"

Chapter 2

As the Robonauts and the Sonic Boomers - along with Heavy Metal and The Best - moved to their assigned sides of the competition field for the final elimination match, "Eye of the Tiger" started playing out the speakers.

"Competing on the red alliance, we have the Robonauts and the Sonic Boomers," announced the MC. "And on the blue alliance, we have The Best teamed up with Heavy Metal." The crowd cheered louder and louder as the MC did his best to talk briefly about each team over and above the music and the noise.

This moment is what every robotics team craved. The adventure and suspense of what would happen after many hours of advance preparation and strategizing. The stakes were real. Only one alliance would win. The excitement and the electricity in the room were palpable!

The field-setters had finished scouring the field, putting each cube in its starting position. Emma would be driving for the Robonauts. It would be Kenzie's job to focus on the activity of all four robots and to help Emma know strategically what needed to happen at each point throughout the match. Corbin would simply stand there cheering the robot on.

Cedar Ridge Academy (CRA), home of The Best, was by far the largest school with the most teams competing at the tournament. It seemed that every time Kenzie turned around, there was yet another student from CRA, all of whom were wearing matching black hoodies with a yellow lightning bolt on the back. The Best's drive team stood in a line across the field from them, looking as confident and as cocky as ever.

"You see that red-headed girl on the left driving for The Best?" asked Emma. "She actually came up to me earlier and said, 'Just wanted to introduce ourselves. We're *The Best.*'"

"Ugh," groaned Kenzie. "Hard to believe she could be so full of herself."

"Yeah. Her name is Nora. And Kyle is their programmer. I'm told that he is exceptionally good at programming. Corbin, don't ever let your skills go to your head like that."

Corbin was also excellent at coding, a powerful asset to the Robonauts. A head-to-head match of skills between these two programmers would be an interesting battle.

"What?" replied Corbin, who wasn't paying attention to the conversation.

"Oh, we're talking about Nora," answered Emma.

"Really? What about Nora?"

"I don't trust her. Anyone that cocky probably wouldn't think twice about pushing the rules and getting away with whatever she could."

Nora happened to make eye contact with the Robonauts from across the field, and she promptly gave them that same big, confident smile and mouthed, *"The Best."*

The three members of the Heavy Metal team were hovering over their robot on their side of the field, doing some kind of last minute work on their robot.

"They better hurry," remarked Corbin. "Looks like the ref is going to start the match any second."

No sooner had Corbin said this, a team member from The Best looked over at the ref and made the time-out symbol with his hands. Indeed, this alliance had a three minute time-out to use. The ref acknowledged the time-out with a single head nod and turned around, making the same time-out symbol to the control table set up behind the field. A moment later, the music had faded out and there was a three minute timer on the screen counting down.

"Hey, Em!" exclaimed Corbin, "did you put fresh rubber bands on Columbia?"

"Of course I did, but..."

As she looked down at Columbia's right intake, she noticed that the rubber band on it was damaged. Not completely snapped, but chipped on the edge and looking like it might snap easily.

"We've gotta get that changed!" said Emma, starting to freak out. "Do either of you have any spares with you?"

Corbin shook his head. Kenzie whipped around to check the countdown timer. 2 minutes, 32 seconds left.

"I've got some in my bag back in the pit. I'll be back with fresh rubber bands ASAP!"

"Make sure to get enough rubber bands for both intakes!" yelled Emma, as Kenzie bolted off towards their pit.

The pit area was outside the gym, across the foyer, down a hall, and in a classroom on the right. And the gym was packed with slow-moving people that kept getting in Kenzie's way and blocking her at every turn. It suddenly became obvious that she would not make it back in time with the rubber bands before the match started. Standing next to the gym door, Kenzie tried to figure out what to do. How could she return to the field for the *final* of all matches without the replacement rubber bands? She looked up at the big screen. 1 minute, 52 seconds left.

"Kenzie?" said a voice behind her.

Kenzie turned around to find another black hoodie with a lightning bolt. How did he know her name?

"Need some of these?" He held out a small clump of fresh, blue rubber bands.

"Um, yes," said Kenzie, a bit stunned that someone from The Best's school was offering help. "Thanks! Gotta go!"

Kenzie grabbed the rubber bands, but before she could get away, the black hooded lightning bolt put his hand on her arm. "I saw you all had trouble with your robot freezing earlier. If you will disconnect your vision sensor and-"

"Look, I really gotta go!" Kenzie interrupted, "Our match is about to start!"

Pulling away, she ran as quickly as possible back towards the field. A short moment later she had returned with 48 seconds left on the clock. She handed the rubber bands to Emma, who immediately started replacing them.

"When did you start getting blue rubber bands?" asked Emma, as she popped the first one in place. She was used to either the regular brown or orange rubber bands.

"Oh, I... I'll explain later. Just focus on getting these on correctly."

Corbin let out a deep breath as Emma got the last rubber band in place.

"And with an entire 14 seconds left on the clock!" said Emma proudly.

The crowd, as excited as ever and growing a bit impatient, began counting down with the clock. First only a few voices saying, "10, 9..."

Then more voices joining in, "8, 7, 6..."

Louder and louder they grew as the countdown continued, "5, 4, 3, 2, 1!"

And then the buzzer sounded. It was time.

Chapter 3

The crowd grew quiet as the MC picked up his mic. "And now the moment we've all been waiting for," he egged on the crowd as the music cranked up loud again. "The winner of this match will be today's Tournament Champion!"

The crowd began cheering wildly as excitement and anticipation once again filled the room.

"Blue team, are you ready?"

Each member of The Best and Heavy Metal gave the required thumbs up sign. Except for Nora, who had to be different by giving her big cocky grin again and nodding her head.

"Red team...." There was an awkward pause. "Would all team members *please* put on your protective eye gear?" The Robonauts and the Sonic Boomers looked at each other.

Corbin laughed sheepishly and pulled his safety glasses down over his eyes. "Sorry, guys," he whispered. Tournament officials always took this safety precaution seriously.

"Red team," the MC started again, "are you ready?"

All three Robonauts and the identical twin Sonic Boomers gave a thumbs up.

"Autonomous begins in 3, 2, 1." The buzzer blasted through the speakers, signaling the start of the match.

Immediately, the four robots began their autonomous routines. The Sonic Boomers successfully pushed one cube into their scoring zone. Although this wasn't a large score by itself, every point was helpful and very much needed.

The Robonauts' routine was more advanced and Corbin had spent many hours getting it to work. He had programmed Columbia to gather three cubes and deliver them in a stack to the scoring zone. The challenging part was completing the stack before autonomous ended - in a very short 15 seconds.

As Columbia moved forward, Corbin bent down on one knee, laser-focused on Columbia, and began cheering for the robot. As each motion of the robot was completed successfully, the Robonauts felt a bit more relieved. Finally, Columbia stacked all three cubes in the scoring zone and backed up. Corbin pulled his fist down, shouting "Yeeeess!" as Kenzie and Emma breathed a sigh of relief.

But while this had been going on, The Best's robot had acquired a collection of six cubes!

"Uh-oh," exclaimed Kenzie when she saw this. "Can they stack that many?"

"We're about to find out," replied Emma.

Heavy Metal had run a routine similar to the Sonic Boomers, pushing one cube into their scoring zone. This meant that the six cubes The Best was attempting to stack would determine the winner of the autonomous period.

The Best's robot was now heading toward the scoring zone. As the robot came to a smooth stop, it began to lay each cube in place, stacking from the bottom up. The cubes rose upward, forming a nice, neat stack. But just before the robot could back away, the buzzer sounded, marking the end of the autonomous period. In order for the stack to count, the robot could not be touching the stack of cubes.

"Did they get it?" shouted Emma.

Nora turned to the crowd in the stands and gave a confident thumbs up, indicating, "Yeah, we got this!"

Other members of Heavy Metal and The Best crowded around the scoring zone with more uncertainty, trying to figure out if the robot was still touching the stack of cubes.

The ref hurried over and got down on his knees next to the robot with a piece of paper. Carefully, he attempted to slide the paper between the stack of cubes and the arms of the robot that were used to deliver the cubes. But it would not fit through.

Standing back up, the ref moved out to where both the crowd and all the teams could see him, and made the symbol for "no good".

Nora had one hand on her hip and did not look happy. Heavy Metal, who had much less of an attitude, stood there disappointed, not knowing how to react.

On the red side of the field, the Sonic Boomers gave each other a high five. Corbin let out a "Take that!" as he looked over at Nora, who was apparently starting to get on his nerves too. Good thing she was turned the other way.

While the ref returned to his position on the back side of the field, the MC took over again, explaining that autonomous had been awarded to the red alliance - a treasured six point bonus. As Kenzie looked over the field, preparing herself for what would come next, she happened to glance up and momentarily lock eyes with Nora. Nora didn't mouth any words to her this time. Rather, she seemed mad. Kenzie flashed a quick half-smile and returned her focus to the field.

Without further introductions, the MC simply announced, "Driver Control in 3, 2, 1," and the robots started moving again. With a collage of colored cubes scattered all over the field, each robot began collecting and stacking cubes in their scoring zones as quickly as possible.

By the time there was roughly 1 minute left in the match, The Best had already scored two tall stacks and were working towards delivering a third stack. Goliath did its best to prevent The Best from completing this third stack by ramming repeatedly into the back of The Best's robot. But despite this valiant attempt, The Best delivered their third stack of cubes successfully. It seemed there was no stopping the blue alliance.

Heavy Metal's robot was now moving around the field, hoarding purple cubes in an attempt to prevent the red alliance from grabbing them. Meanwhile, the red alliance had a much smaller showing of cubes in their scoring zones.

Kenzie quickly counted all the cubes and colors in the towers and scoring zones. Then, she entered these numbers into VRC

Hub, the VEX scoring app on her phone. Although the blue alliance had more cubes stacked than the red alliance, little Goliath had managed to put two purple cubes in the short towers. Cubes in towers were extremely valuable because they functioned as multipliers for all other stacked cubes of the same color, dramatically increasing the teams' score. Indeed, the color of *all* the cubes stacked by the red alliance was purple, just as planned. Not good for the blue alliance who only had one purple cube in all of their stacks. With all the numbers entered and with the autonomous bonus figured in - and if Kenzie had accurately tallied the score - it looked like the blue alliance was ahead by six points. This was not a big surprise considering how good these teams were.

Emma was driving around the field collecting as many purple cubes as she could. She already had three but was trying to pick up a fourth cube before making her way to the scoring zone to deliver them.

Although Corbin had yet to figure out the details, he had been working on a plan over the past few weeks that would help with their strategy of focusing on a single color of cube. Mounted on the top of Columbia, the VEX vision sensor had the ability to function as eyes for the robot, identifying the color of each cube that came into the robot's field of vision. Using measurements from the vision sensor and a bit of math, the robot could be programmed to center itself in front of a cube. This would make it much easier for Emma to pick up the cube quickly and at the correct orientation.

"How are we doing?" asked Emma, retaining her focus as she picked up the fourth purple cube.

"Well, we're putting up a good fight," replied Kenzie, "but-"

"No waaaay!" blurted out Corbin. He was pointing across the field at Heavy Metal's robot, which was sitting right next to the blue alliance's rectangular scoring zone, with a pile of cubes from two collapsed towers tumbled all around it.

"What happened?" yelled Kenzie.

"Heavy Metal wasn't paying close attention!" Corbin said excitedly. "When they turned the robot around, the back end of their tray swung right into their stacks, knocking them down!"

The distraught look on the faces of both teams of the blue alliance confirmed that this was exactly what had happened, along with a loud and sharp "What on earth?!" coming from Nora.

There were now only 28 seconds left in the match and the clock was counting down quickly. There would not be enough time for The Best to collect more cubes and restore their massive double towers. With the few purple cubes that the red alliance already had in place and the two purple cubes serving as multipliers in the towers, courtesy of little Goliath, did the red alliance actually have a chance of winning? Kenzie hurriedly updated the tally of the score on her phone - and her jaw dropped when she saw the results.

"Emma! You *have* to stack those cubes! If you can get them, I think we're gonna win this thing!"

With 18 seconds left on the clock, it was very possible that Emma could successfully complete this stack in time. In her previous practice attempts, she had consistently delivered four cubes in anywhere from 10 to 15 seconds. Emma carefully brought Columbia to a stop in front of the scoring zone and every eye in the room watched closely as the stack of four cubes started to tilt forward.

Suddenly, the robot stopped moving, frozen in place.

Emma began to panic. She started punching buttons on the controller harder, but to no avail. Columbia had gone unresponsive. Before anyone could figure out what had happened, the buzzer went off, signaling the end of the match.

The red alliance stood there, stunned. Why did this problem have to happen *during the final*? And why only to Columbia?

The first person to break the lull was Nora, who screeched "Yeeees!" Following Nora, all the black hoodies with the lightning bolts in the room screamed out in victory along with everyone from team Heavy Metal.

Winning the final match at a robotics tournament was exhilarating. And losing was equally disappointing. Even though the Robonauts had done very well, this loss was a huge letdown. Especially with the way it had happened.

As the winning alliance gave high fives, bumped fists, and shouted in victory, Kenzie, Emma, and Corbin slowly made their way back to the bleachers with their robot. Still in a bit of shock over it all, they sat down on the front row to wait for the awards ceremony to begin.

Chapter 4

"The awards ceremony will begin in about 15 minutes."

The MC then proceeded to thank all the teams for their efforts and make sure everyone knew that left-over pizza was available at the concession stand. But Corbin was in his own world, tuning everything out.

"How could our robot simply freeze?" he thought to himself. Slipping his laptop out of his backpack, he scrolled through the pages of code he had written to control the robot. Surely there was something he had missed, something that was causing Columbia to come to a complete standstill.

"Find anything?" asked Emma, looking over onto the laptop screen.

"Nope, nothing at all," replied Corbin in frustration. "Although, you know how button Y stops the robot at just the right distance from the corner when Columbia is headed towards the scoring zone?"

"Yeeees?" replied Emma curiously.

"When you press button Y," explained Corbin, "a loop runs that continuously checks to see if the robot has made it to the correct stopping point. What if that loop isn't completing and is running endlessly?"

"Would that cause the robot to freeze up?" asked Emma.

"It's my best theory at the moment," said Corbin as he looked over his code again.

“Only one problem with that theory,” said Kenzie, “Columbia didn’t freeze until after the stack was being delivered.”

Kenzie was right.

As the team sat there trying to figure out what had happened, Coach Warren climbed up into the bleachers and sat on the row behind them. “Pizza anyone?” He opened a box of four-hour-old pizza that he had picked up from the concession stand.

“Yes, sir!” said Corbin, without hesitation. He closed the screen on his laptop and grabbed a couple of slices. The only thing Corbin liked better than programming was food. It had been a long day with no time for lunch. Emma and Kenzie both politely declined any pizza - when what they were really thinking was, *Really? Cold pizza? Ewww!*

Coach Warren was a young science teacher at Valley Creek High School. His job kept him busy throughout the year, meaning that the Robonauts were almost completely student-driven. Still, he made time to drive the group to tournaments, beg the school administration for funding, and was always as supportive of the team as he could be in every way possible. Including bringing the team cold pizza at the end of a hard tournament.

“Did you see Columbia freeze up in our final match?” asked Corbin, with his mouth full.

“Yes. I’m so sorry about that! I was able to get a video of it,” said Coach Warren, as he pulled his phone out of his pocket. But before they could stop to look at the video, the MC picked up the mic and started the awards ceremony.

As the crowd settled down and each award was presented, everything slowly became a blur for the Robonauts. Indeed, it had been a hard and frustrating day, and the team was completely worn out. After the awards ceremony, after packing up all their equipment from the pit, and after piling everything in the back of their school van, Coach Warren pulled out of the parking lot and began the team’s long drive home. Ten minutes into the drive Corbin was sound asleep, leaning against the side window in the front seat. Kenzie and Emma were in the back seat with their eyelids getting heavy, well on their way to drifting off to sleep too.

Looking down in the seat to her left, Kenzie noticed a blue rubber band that had fallen out of her pocket. She picked it up and stretched it a bit, thinking back to her encounter with the guy from CRA who had unexpectedly helped them out. Why had he chosen to help? And what was he trying to say about the vision sensor?

One of these days we are going to beat The Best at a tournament, thought Kenzie. She wasn't quite sure how this would happen, but she was becoming more and more resolved to show that team once and for all that other teams could also be the best. This was the last thought that passed through Kenzie's mind before her eyes closed and she fell fast asleep.

Chapter 5

Each tournament is a peak of the overall robotics adventure - but this is not where most of the work occurs. Rather, the hours of building, programming, journaling, strategizing, etc. happen elsewhere and beforehand. Preparation may involve many hours on a practice field. Or in a lab where the robot is prototyped and built. Or late at night in the basement of a team member's house. Or while browsing the VEX Forum over lunch. Or, in simply thinking things through - and one day coming up with an idea that you know will work well (even if you have to wait a few days to try it when your robotics team is able to meet again). As is true in many aspects of life, the more you put into robotics, the more you will get out of it.

* * *

"So, you can find the third side of the triangle using the Pythagorean theorem." The lecture continued to drone on in the background.

Kenzie was only somewhat paying attention. She glanced up at the clock. Five more minutes? Ugh. It seemed like this class would never end. At least it was the last class of the day. When would she ever use that pytha... something-or-another? Math was not her subject. Kenzie stared out the window, twirling her long brown hair as she often did when she was deep in thought.

Since the catastrophic ending of the final match at Tumultuous Towers, all Kenzie could think about was figuring out the problem with the robot and getting things fixed.

"Kenzie, what did you get for the hypotenuse?" said Mr. Brown, her math teacher, interrupting her thoughts.

Jumping back into the present, Kenzie realized that she didn't even know which problem the class was on.

"Umm... I'm not sure," she mumbled.

Emma, who sat two seats up and one row over, tilted her dark black hair to the side and glanced back at her friend through her blue-framed glasses. With a smile on her face, Emma had a knack for finding humor in most anything.

From the front row, Gordon's hand shot up in the air. "I know!" he whispered under his breath. Gordon was good at math and always knew the answer.

"Yes, Gordon?" replied Mr. Brown with a hint of reluctance and possibly a slight eye roll.

At least Kenzie was no longer on the hot seat. She would have to thank Gordon later for this.

Kenzie glanced at the clock again. Two more minutes until the bell.

As the class slooowly came to a close, the evening's math assignment was made. *When am I going to have time for thirty math problems?* thought Kenzie. But she would have to figure this out later. At the moment, there were bigger issues to deal with. Her plan was to make a quick stop by her locker to deposit all of her books and then go plant herself in the school robotics lab for the rest of the afternoon. There she could make good use of her time making progress towards fixing the "freezing problem" that was plaguing the Robonauts.

Within moments after the bell rang, students were filing out of the classroom, and the hallways filled with voices and the clanging of lockers opening and closing.

As Kenzie shoved her books up into the top section of her locker, she heard some faint giggling from several lockers away.

"What's up?"

"Oh, nothing," replied Emma, as she tried to contain herself.

Shrugging her shoulders, Kenzie turned back to her locker. Then she saw it. Posted inside the door of Kenzie's locker, there was normally a huge sticker that proudly displayed "I ♥ Robotics". But someone – and there was no doubt who – had taped up a piece of paper over the word "Robotics" so that it now read, "I ♥ Math".

It took all of two seconds for the sign to be ripped down, wadded up, and thrown down the hall in Emma's general direction.

"Ha ha," barked Kenzie.

Emma peeked around the edge of her locker door and smiled. "Ready to go work with Columbia?"

"Oh yes. Anything other than math class."

Chapter 6

Kenzie and Emma had been best friends since they both joined the Valley Creek robotics team the previous school year. While both girls loved the adventure of robotics, they were as different as night and day. Kenzie was always coming up with new ideas for making Columbia better and more efficient. But Emma would be the first to point out that all such ideas have to be workable. And, of course, any needed parts have to be available. And those parts cost money, which the school likely didn't have. Indeed, Emma was the logical one who looked at things more realistically. Kenzie's creativity and push-forward attitude combined with Emma's voice of reason and realism always made for an interesting combination.

"So, I've been thinking about Columbia and what could have made it freeze up at Tumultuous Towers," said Kenzie, as the two girls began making their way down the hall to the robotics lab. "ESD."

"ESD?" asked Emma. "What's that? Sounds like some horrible disease."

"I found it on the VEX Forum. It stands for electrostatic discharge. Apparently, friction-induced static buildup can damage the robot's brain and interfere with sensors connected to it."

"You know, maybe ESD is your problem in Mr. Brown's class," joked Emma. "All you need to do is rid yourself of static and you'll suddenly be a mathematical genius!"

Kenzie smirked. She could always count on Emma's jokes and slightly sarcastic comments to lighten the mood.

The two girls walked down the hall, around the corner, and into the robotics lab. They found Columbia sitting on the practice field in front of the scoring zone in the same position that it had frozen over the weekend. And there was Corbin, with his laptop opened, working hard to figure out the problem. The back of Corbin's shirt appropriately read, "It should work IN THEORY."

The robotics lab at Valley Creek High wasn't elaborate. Just half of a competition field crammed into one side of a classroom with bunches of desks and chairs scooted over to the other side. The Robonauts' work area consisted of two tables positioned under a long whiteboard at the front of the room. Plastic bins were stored under the tables to hold parts, tools, etc. The back side of the room had a wall of windows, providing a full view of the school yard and parking lot.

"Got everything figured out?" asked Emma, grabbing a wrench to fiddle with as she joined Corbin in staring at the robot.

"Not hardly," replied Corbin in frustration. "I've been through our code a million times and I can't find any problems with it." An outline of Corbin's code was freshly written on the white board where he had been tracing the flow and thinking through the logic, all to no avail.

Kenzie rolled out a swivel chair from under one of the work tables, sat down in it, and wheeled herself over to her teammates on the field. "Have you considered static interference?"

"Yeah," said Corbin, "I read about it on the VEX Forum, but there isn't really a way to know for sure if this is the source of the problem."

"Even if ESD is causing Columbia to freeze, what could we do to prevent it?" asked Emma.

Neither Kenzie nor Corbin had an answer to this.

"Well, could the problem be a connection issue? Maybe the cable from the radio to the brain is faulty and intermittently loses communication with the controller? Or maybe the controller itself is the problem?" suggested Kenzie, searching for a solution.

"Possibly," said Emma. "But to test this theory we would need to purchase replacement cables along with a new controller. Why not throw in a new brain while we're at it? And we can sell cookies for the next six months to pay for it!"

Point made. There was no money or time for this. New equipment was not an option.

Corbin's head popped up. "Cookies?"

"Let me think on this," replied Kenzie, ignoring Corbin's comment and slowly twirling her hair in thought.

Emma tossed her wrench to the side and let out a sigh.

This scene was all too familiar. The entire team sitting around the robot with more problems than solutions, not really knowing what to do. No question about it, robotics was hard.

Other than Corbin's slow sipping from a huge bottle of Dr. Pepper which he always seemed to have with him, the Robonauts sat there in silence for a minute or so, trying to figure out what to do. Finally Emma broke the silence. "Well, at least we know the problem isn't with the mechanism that tilts the tray to stack the cubes. Last night I tried replacing the gears on it, just in case anything was getting jammed up."

Corbin shot her a look. "You what?"

"I changed the gear ratio so we would have more power when lifting up the tray of cubes. Figured we ought to rule out overworking the motor as the problem."

Leaning in to examine the back of the robot, he then saw what Emma had done.

"Umm... Thanks for doing that," said Corbin, unconvincingly. "But with all the metal in place to hold the extra gears, that adds a lot of weight to the back of the robot."

"Yes, and I was thinking this could also help keep Columbia from getting pushed around as easily on the field," replied Emma, who wasn't yet sensing or understanding Corbin's growing frustration.

Determined to stay out of the pending disagreement, Kenzie remained quiet, slowing spinning herself around in her chair.

"You do realize," said Corbin sternly, "that when you make major changes like this, I have to go through and re-program significant portions of our code?"

The expression on Emma's face suddenly went blank. "Sorry, I didn't realize..." she said softly. "But how does changing a gear structure make you have to reprogram the robot?"

Although Emma meant well, the relationship between a builder and a programmer can be fragile. Especially when unexpected changes by the builder require the code to be updated. But changes to the code can just as easily lead to more work for the builder. So good communication is essential.

"Any physical changes to the robot can affect the programming," explained Corbin, "particularly if the robot's size, shape, or weight are altered significantly. For example, if the robot gets heavier, the motors that control the wheels may have to be reprogrammed to use greater force when moving forward or turning. Or if-"

"Hey!" Kenzie interrupted Corbin in mid-sentence. "I completely forgot to tell you all."

"Yes?" said Emma, thankful that Kenzie was bringing an end to Corbin's lecture.

"Remember when I went to get rubber bands before the final match at Tumultuous Towers? Well, I never told you where I got those. Once I realized that I didn't have enough time to make it to the pit and bring back our rubber bands, I wasn't sure what to do. That's when a guy caught me and offered to share his rubber bands with us."

Sharing parts or tools between teams isn't uncommon or surprising at robotics tournaments. At least it hadn't been unusual for the Robonauts. That is, until Kenzie described who it was that had helped them.

"The rubber bands we used," continued Kenzie, "came from CRA - from one of the students wearing a black hoodie with the lightning bolt. I don't know the guy's name or why he took time to find me and help us during the time-out."

"Maybe the rubber bands were defective and sent by Nora, hoping they would mess us up," suggested Corbin sarcastically. "Nora had it out for us all day long and I wouldn't put anything past her."

Emma nodded her head in agreement, hoping she was no longer the main object of Corbin's frustration.

“No,” said Kenzie thoughtfully. “I think he really wanted to help. Something about the way he approached me was different. He actually mentioned the freezing problem that had been going on all day and said something about disconnecting our vision sensor to prevent this.”

“Do we even use the vision sensor for anything?” asked Emma.

“We use it minimally,” replied Corbin. “We run it constantly to help identify the color of any cube that comes in front of the robot. But honestly, we don’t do anything with this information yet because it’s faster and more efficient for our team to look at the cubes ourselves and decide on the spot if we want to pick each one up or not. But, check this out.”

Corbin picked up the controller and pressed a few buttons to start a program running. Then, picking up a purple cube, he held it out in front of Columbia and moved it back and forth from left to right. As he moved the cube, Columbia “saw” it with the vision sensor and adjusted its position to stay facing the cube as it moved.

“Pretty neat, isn’t it?” remarked Corbin. “I just haven’t figured out a good way to use this in a match.”

“But what would the vision sensor have to do with the freezing problem?” asked Kenzie.

“I don’t know,” admitted Corbin. “It sure is hard to fix a problem when you can’t duplicate it.”

“I have an idea!” said Emma brightly, completely changing the subject. “Maybe I could try to mount the brain a bit higher to get it further away from the static electricity created by the wheels? All I would have to do,” she continued as she got more excited about her idea, “is rearrange a few parts like-”

Emma suddenly paused as she saw Corbin shaking his head no.

Again, there was silence as the team realized they were no closer to solving this problem. The only thing accomplished was that Corbin now had to make bunches of updates to the programming. And Kenzie would now have to stay up late updating the journal in addition to doing the thirty assigned math problems. It would be a very late night.

Glancing out the window, the team watched a navy blue SUV pull into the pickup circle by the front entrance.

“There’s my parents,” said Corbin, as he shut down his laptop, threw it in his backpack, and started out the door with his Dr. Pepper.

“Oh, I meant to ask,” Corbin said, turning around. “Did either of you get the video from Coach Warren of the final match?”

Both girls shook their heads no.

Normally Coach Warren was around for these after-school team meetings, but he hadn’t been seen today since school let out.

“Hopefully we can get all the videos soon. We’ll get this problem figured out eventually.” With this, Corbin disappeared into the hallway to head home for the day.

There are times when a team needs to step away from the problems they are trying to solve just to maintain their sanity - and, for the Robonauts, this was one of those times.

Chapter 7

Emma's house was a short walk from Valley Creek High and Kenzie's house was only a few minutes further down the same street. This made it convenient for both girls to stay late for robotics most any afternoon they wanted.

"Wanna stop by and work on thirty math problems with me tonight?" offered Emma.

"No, not really," answered Kenzie honestly. "But I guess we need to. Maybe we can have a Pytha-whatever party."

"That would be Pythagorean."

"Speaking of which, where is my math book?" asked Kenzie, searching around the room. "I think I must have left it in my locker."

"Trying to get out of doing math, huh?"

"I'll stop by my locker to get it," said Kenzie, ignoring her comment. "Meet me out front at the flagpole in two minutes?"

Emma nodded her head affirmatively and Kenzie began making her way back down the empty hallway, around the corner, and to her locker. The thought of staying up late doing math homework made her stomach turn, especially when there was so much robotics work to do.

Reaching into the top section of her locker, Kenzie paused as she heard voices coming from an office down the hall. Although the voices were talking quietly, the words carried clearly into the hallway. She didn't want to eavesdrop, but her ears perked up when she heard the word *robotics*, and she couldn't resist. Kenzie recognized one of the voices as that of Coach Warren, but the other voice was unfamiliar. Pulling her math book out of her locker, Kenzie begin to listen more intently to what was being said. And she began to smile. She couldn't wait to tell Emma and Corbin about this! Or maybe she should save this bit of information and surprise them with it later?

As the conversation continued, it became obvious that Coach Warren and his guest were wrapping things up and were about to enter the hallway. Quickly, Kenzie closed her locker and started walking back the way she had come. As she rounded the corner, she took one last glimpse back and saw Coach Warren and his guest emerging from the office. It was Curly from Tumultuous Towers! His long, curly hair and red cap made him unmistakable! Kenzie hurried away to the flagpole to meet up with Emma.

On the way home, they talked all about Tumultuous Towers, about their classes, and how Gordon had completely saved Kenzie in math class earlier in the day. But Kenzie decided not to share the news she had heard. At least, not yet. She would wait for just the right moment!

* * *

Scattered across the kitchen table was an empty popcorn bowl, a bag of half-eaten chips, and several mini-size candy bar wrappers. The large clock over the kitchen sink read 10:34pm. And sitting at the end of the table with her head facedown on her math book was Kenzie. Emma was working on the floor, sitting cross-legged and finishing up her last math problem.

"How's it going up there?" asked Emma, primarily concerned that Kenzie was still awake and making progress.

No answer.

Just as I suspected, thought Emma to herself.

Before Emma could get up to check on her, Kenzie's phone went off. It was on the table right next to her head, blasting the first few measures of "Eye of the Tiger" directly into her ear.

Startled, Kenzie sat up, grabbed her phone and shut off the sound. Hopefully this hadn't woken up Emma's parents, who were asleep at the other end of the house.

"Hey, sleepy head, hope you're making a good dent in your homework up there."

"Yeah, I'm working on it," said Kenzie with an enormous yawn. The math wasn't nearly as hard as she had thought it would be, and she was mostly done. She could finish up the last couple of problems before school tomorrow.

Again, "Eye of the Tiger" sounded - fortunately this time not quite as loud.

"That would be Corbin," said Emma, checking her phone. "Looks like he's texting both of us."

Hey U 2, the first text read. *Coach finally sent us the videos from the comp. Check out these matches. Notice anything about them?* Each video slowly loaded for the girls to look at.

Emma moved up to the kitchen table with Kenzie and they started looking through the videos one by one. All these videos did have something thing in common. Each of these matches were ones in which a robot had frozen on the field, a gloomy reminder of the problems that had plagued many teams at the tournament.

Before the girls could finish the last video, both of their phones went off again, this time a FaceTime call from Corbin.

"Umm... any idea what time it is?" said Emma after answering the call.

"Yeah, yeah, I know. But what do you all think about this?" asked Corbin.

"About what?" asked Kenzie. "The robots freezing?"

"No, about the videos!" continued Corbin. "Did you all notice who was on the field competing every time a robot froze?"

Kenzie and Emma glanced at each other and shrugged their shoulders.

Corbin leaned forward. "The Best!"

"So, what does that mean?" asked Kenzie, who was still confused.

Corbin raised his eyebrows. "Of all the matches where a robot froze at the tournament, what are the chances that any one team would just *happen* to be on the field competing every single time?"

Emma was already double-checking the videos. "In the first video, The Best. In the second video, The Best...." There was a definite pattern. The Best were always there.

Kenzie could feel her blood start to boil. "I smell a rat."

"Do you all really think Nora's team could be involved in the freezing problem?" asked Emma.

"It appears that way," answered Corbin. "It wouldn't surprise me if they caused it."

"Each time a robot froze, it did directly benefit Nora's team, helping them get all the way to the final," Kenzie noted. "If they were able to stop their opponents' robots in their tracks, it's no wonder they did so well."

"Well, for the moment, we don't have any proof and I don't know what we can do about it. But whatever they are doing, we're going to find out," Corbin declared with a firm resolve.

"Agreed," said Emma. Kenzie nodded her head firmly.

There was a pause.

"It's late, and I need to get some sleep," said Corbin. "By the way, Kenzie, nice hairdo." The front of her hair was nicely flattened, following a curiously similar contour to the open math book on the table. Before she could respond with a smart remark, Corbin left the call. Probably in his best interest.

Kenzie and Emma looked at each other and burst out laughing. Not that having math book hair was terribly comical, but they were so worn out that everything was starting to seem funny, and it was past time for Kenzie to get home.

Leaving Emma's house, Kenzie started her walk home, thinking through all that had happened in robotics throughout the day. Robotics has a way of taking over your time, your energy, and your thoughts. But Kenzie wouldn't trade any of it for the world. It gave her something to pour herself into, have fun with, and learn valuable skills all at the same time.

About halfway to her house, Kenzie's phone went off again. This time it was a text from Gordon. It just said, *Hey, this is Gordon.* What was he doing texting her this late? In fact, what was he doing texting her at all? How did he get her number? It's not like she was good friends with him.

How's the math coming? continued Gordon.

Fine, only a couple of problems left, Kenzie replied. *BTW, thx for taking the heat off of me in class today.*

No prob, replied Gordon.

Ok, well, good night. Kenzie texted, awkwardly trying to end the conversation.

Good night! Let me know if u need any help with your final math problems in the morning.

After sending one last thumbs up to Gordon, "Eye of the Tiger" started playing yet again. This time it was from Emma. *Forgot to tell you. Gordon asked me for your number earlier today. He obviously likes u. Don't break his heart too harshly!*

Ha ha, replied Kenzie.

Emma replied with a long string of smiley face and heart emojis.

And that was the end of the conversation.

Chapter 8

A team like The Best can make for a good story. But in reality, VEX Robotics teams are nothing like this at all. Teams must follow the rules of the game and the Code of Conduct required by REC Foundation, or risk being disqualified. Yes, the competition and desire to win is intense. Yes, there can be drama. But healthy competition makes everyone involved continually strive to be better. When a team does not do well at a tournament, more often than not, that team returns to future competitions, fighting harder with more experience and knowledge than before.

* * *

Two days prior to Tumultuous Towers

Talk about crushing the competition! Kyle said to himself. He couldn't believe how well his plan was working. He smugly looked down at his robot, appropriately named Destroyer. It was living up to its name.

Kyle put Destroyer back on the practice field for another trial run. Then walking over to the table beside the field, he made a few changes to the code on his laptop. Finally, he signaled to his partner in crime to go ahead with the next run.

As the robot began the well-practiced routine of collecting several cubes and heading towards the scoring zone to deliver them, Kyle's red-headed teammate waited for the right moment. As soon as the robot was about to deliver the cubes, Nora pressed button A on the controller. Immediately the robot froze in its tracks. A glimpse of what would happen to any team who got in their way.

"You are an evil genius." said Nora, calmly and smoothly.

"Oh, you haven't seen anything yet," bragged Kyle. "If my latest idea works as well as I think it will, our team will be completely unstoppable."

"Oh, can you make parts fall off of our opponent's robot?"

Nora obviously did not understand robotics as well as she liked to make everyone think that she did.

"Uh, no. But I think we can do even worse than that," continued Kyle.

Nora raised her eyebrows. "Really? What could be worse than a robot falling apart on the field?"

"How about taking the robot out of commission all together?" said Kyle deviously. "You see, when we freeze a robot, it can be fixed by simply restarting the brain. But what if restarting wouldn't fix it?"

Nora liked the sound of this, but doubted that this programming genius could pull off something this dramatic.

"How are you gonna make this happen?" asked Nora doubtfully.

"Well, it's untested, but I think it has an excellent chance of working," replied Kyle.

"So, let's test it," continued Nora, getting a bit impatient.

"Noooo, we can't test it."

"Why not?" demanded Nora.

"Do you really want to test out a plan to damage a robot," Kyle paused, "on *our* robot?"

Nora was starting to realize that Kyle was serious about this.

"If this works," Kyle continued proudly, "our opposing alliances won't know what hit them."

Feeling the eyes of someone on her, Nora turned around to see Alex standing in the doorway.

"How long have you been standing there? How much did you hear?"

Not knowing what to say, Alex just stood there silent.

Alex was in his second year of robotics at Cedar Ridge Academy (CRA) on a team named the Hobbits, a sister team to the Best. While not being an expert in any one area of robotics, Alex was an asset to his team because of his willingness to jump in and learn, and help out in any way that he could. More importantly, he had a good dose of common sense. And at this moment in time, his common sense was telling him that he had just walked into the *wrong* conversation.

Nora made her way over to Alex, pulled him into the room and slammed the door. Pointing at one of the seats in the classroom, the expression on Nora's face said, *Sit down!*.

Alex slowly sat down, not sure what was coming next.

"OK," Nora began, "Here is how this is going to work. You didn't hear anything. You didn't see anything. You know nothing!"

Kyle looked carefully over at Alex to see how he would react.

"Life will go on as if you never came into this room today. Is that understood?" Nora said forcefully.

Mustering up a bit of courage, Alex made eye contact with Nora.

"Why do you need to cheat?" he asked softly. "Your team has a good robot, and you can do well without messing with other robots."

"Because we *can*," answered Kyle from across the room. Indeed, much of Kyle's motivation was simply the power that came from being able to control things through hacking.

"And because we not only want to win," added Nora, "but we want to *dominate*. And we *will* succeed!"

Alex sat there stunned, not believing what he was hearing. But apparently Nora and Kyle were dead serious.

"If you even *think* about breathing a word of this to anyone, your team's robot will be the first one we attack," warned Kyle. Nora raised her eyebrows as an exclamation point to their threat.

With only two days before Tumultuous Towers, the last thing the Hobbits needed was a damaged robot.

"I think you are bluffing," said Alex. "How is it even possible to freeze or damage another robot through coding?"

One of Kyle's inherent weaknesses was that he liked to hear himself talk, taking advantage of every opportunity possible to brag about what he could do. After all, what was the fun in being "the best" at something if no one knew about the brains behind it all? *Now that he knows about our plans,* Kyle reasoned to himself, *what could it hurt to explain it to him? Maybe it would scare him into keeping his mouth shut.*

"Here is how it works," said Kyle in a tone of superiority. "The brain of each robot stores and executes programs, just like any computer does - but the brain also has the same inherent weaknesses of a computer. When the subroutine of a program executes, it uses a certain amount of memory. But I've discovered a flaw in the brain. When using a sensor, the memory utilized by the subroutines that reference it is never fully released. If this happens over and over, the amount of available memory continually decreases until there isn't enough left, making the robot freeze.

"And let's just say," Kyle boasted, "that I have figured out a way to identify other robots over the field's network and, with the push of a button, speed up this process."

"In other words," interrupted Nora, who was growing a bit impatient with Kyle's self-indulging and overly technical monologue, "I can immediately freeze any opposing robot on the field with a push of button A or B - depending on which robot I choose to zap."

"Just wait until you see what the X and Y buttons do," added Kyle, as if he had already seen it work. "We can redirect battery current from Destroyer so that it travels over the field's network and overloads the circuits-"

"And destroys the robot's brain," Nora cut in again.

"If you mess with our robot, or any other team's robot, you will be sorry you did," said Alex timidly, not having anything to back up this threat.

Nora rolled her eyes. "If that's the way you feel, I guess we'll just have to practice on your robot," announced Nora.

Alex watched in dismay as Nora walked across the room to the Shire, the cabinet where the Hobbits' robot, Frodo, was stored. She flung open the door, yanked the robot out, and marched it over to the opposite end of the field from Destroyer. As she connected the robot's controller to the field system, Kyle started up Tournament Manager, preparing to begin a completely unfair match against the defenseless Frodo.

"Please don't," begged Alex, "our team has spent hours and hours working on Frodo!"

As Alex said this, Nora stopped. Glancing at Kyle with an evil grin, she got up and walked backed over to Alex.

"If you open your mouth to anyone, anyone at all, we will destroy your robot! If you even *think* about telling anyone about this, we *will* destroy your robot! Understand?"

With a sigh of resignation, Alex nodded his head.

"Now out!" said Nora, as she sternly pointed to the door.

Without hesitating, Alex got up and dashed out of the room.

"Think he'll keep his mouth shut?" asked Kyle.

"Yeah, because he's scared we're gonna fry his team's silly robot. Anyway, we have more important things to do than worry about him. Like figuring out who our main competition is gonna be at Tumultuous Towers and which robots we need to target."

Kyle pulled up robotevents.com on his laptop and began scanning through all the teams that would be at the upcoming tournament. One team in particular caught his eye. Doing a quick lookup, he found the team's YouTube channel. "What do we know about this team from Valley Creek High?"

"Never heard of them," said Nora. "Are they any good?"

"Nah, we can beat them," replied Kyle, reluctant to admit that any team could ever rival The Best, "but if any team at this competition has a chance of catching us on a bad day, I think it would be them. It looks like their team is called the Robonauts. Three members, a guy named Corbin and two girls, Emma and Kenzie."

Kyle would never understand why robotics teams would post so much information about themselves online. And why would anyone post detailed videos of their robot? (The thought never crossed his mind that doing so might help other teams, particularly new teams who are learning.) Right here was a video of Corbin explaining how they hoped to use their vision sensor in a tournament. Kyle would never post something like this himself. He couldn't stand the thought of anyone stealing his ideas. But he was glad the Robonauts posted their information. This made it all the easier for Kyle to devise a plan of attack.

"Looks like we'll just have to exploit their vision sensor," remarked Kyle, "since they like to use it so much."

As Kyle and Nora continued scheming, Alex wrote "Robonauts" on the palm of his hand. He was sitting on the floor outside the robotics lab, listening to all that was being said. As loud as Nora and Kyle talked, it was easy to hear them plotting from the hallway.

Alex could always go to his teachers and try to explain what was happening. But Nora and Kyle would deny it. And how would his teachers be able to help anyway? Surely there was some way he could fight this without having the wrath of The Best fall on his team and their robot.

Not wanting to be around when Kyle and Nora exited the lab, Alex decided it was time - past time - to get out of there. Now he would go home and see what he could find out about the Robonauts.

Chapter 9

The day of Tumultuous Towers

"Did you pack all the extras you usually keep up with at competitions? Like rubber bands and cable ties?" asked Cory.

Alex, only somewhat paying attention, nodded his head and continued to look out the bus window. Nora, Kyle, and Eugene, the three members of The Best, were sitting several rows behind them in the back seats of the bus. Alex didn't know much about Eugene other than that he was their builder and everyone called him Euggie.

"Are you alright?" persisted Cory, hoping for more of an answer, "because we *are* competing today and I sure *hope* you are doing OK."

"I'm good," lied Alex, in his usual soft-toned voice. The truth was, he felt like Nora and Kyle were silently watching and were ready to pounce on him if he said or did anything to counter their mischievous plans for the day.

"You know that after we get off the bus, this day will move fast and I'm going to need your head in the game."

Cory, Alex's friend and the second half of the Hobbits, always had a way of being matter-of-fact and cutting straight to the point. This, along with her endless supply of energy, made her a valuable asset to the team. On tournament days, Cory was always observing, analyzing the situation, and thinking through what needed to happen next. At this moment, her focus was on Alex and whatever was going on behind his blank stare. With a determination to figure out what was going on, she looked directly at him.

"C'mon, out with it," said Cory.

Maybe it *would* be good to talk to someone.

"Can you keep a secret?" whispered Alex.

"Um, sure?"

"And you promise you won't freak out?"

Now Cory was getting curious. "OK, I promise. Now out with it."

Sliding down in his seat and keeping his voice low so that no one else could hear, Alex began to tell Cory about his untimely run-in with Nora and Kyle.

"They *what*?" blurted Cory, quite loudly.

Suddenly realizing how loud she was, she sunk down a bit lower in her seat and asked again, this time much softer, but still quite pointedly. "They did what?"

Alex told Cory everything he knew about The Best's plan to cheat, to freeze robots at today's tournament, and how they were working towards destroying robots. Worst of all, he told her about their threat to destroy Frodo if he told anyone.

At first Cory just sat there absorbing all of this information. The thought of Nora and Kyle destroying Frodo after all the hours she had put into building that robot made her nauseous. But she was a builder and a fixer. And this was a problem that needed to be fixed.

"So, what do you think we ought to do?" asked Alex.

"Well, for starters," said Cory, "I think we need to watch them today and learn all we can about what they are doing. Who knows, maybe we can figure out a way to turn their cheating around on them?"

If anyone had the energy and drive for this, it would be Cory.

The bus finally arrived at Tumultuous Towers and pulled into the parking lot. As soon as it came to a stop, Coach Blakely, the robotics coach at CRA, stood up at the front.

"Listen up," she shouted so everyone could hear. "Today's competition is in the gym. I'm passing out a map to help you find inspection, registration, your team's pit area, and the Skills field."

"As you know," continued Coach Blakely, "it's important to get through inspection as soon as you can. The last thing you want is for an inspector to find some problem with your robot at the last minute when there may not be enough time to get it fixed."

Alex was accustomed to hearing this speech at the start of every robotics competition. Nevertheless, it was always a good reminder.

Coach Blakely admonished all her teams to follow the schedule closely, to get to the queuing tables early for each of their matches, and to keep a clean pit. She reminded them to always be prepared for judges to come around for interviews, to make sure to do their Skills runs, and to take time to scout out other teams for potential alliances. Then Coach Blakely ended with a simple, "OK, let's unload!"

The bus became a mad house with everyone gathering their things. With some careful maneuvering, Alex and Cory were able to maintain their distance from The Best as they worked their way off the bus and unloaded their equipment from the cargo bay. After gathering the wooden crate that contained Frodo, Alex began rolling it towards the gym. Cory grabbed the rope on the front of the crate to help guide it in the right direction.

Although The Best and their antics were weighing heavily on Alex's mind, it helped that Cory was now aware of the situation. Still, it was going to be a long day.

Chapter 10

After losing their first two qualification matches, Alex and Cory had a break before their next match and it was a good time for them to do their Skills runs. Each team was allowed up to three driver-controlled runs and another three autonomous runs. Cory excelled at controlling Frodo in driver-control, probably because she had spent many hours practicing and mapping out the path the robot would follow. In fact, Driver Skills seemed to be the Hobbits' greatest strength as they never seemed to do terribly well competing head-to-head with other robots in the tournament. Their not-so-good start in today's competition was evidence of this. But every team has their strengths and weaknesses. Although neither Alex nor Cory were experts in coding, they had programmed Frodo to perform a short autonomous routine to place two cubes in towers and deliver a stack of three cubes into a scoring zone.

When Alex and Cory entered the room where the Skills field was set up, they found one other team that was finishing up, the field operator, two field-setters, and a handful of students scattered around the room observing the strategies that other teams were using. And sitting near the Skills field was Euggie.

"What's Euggie doing here?" whispered Cory. "Why is he not in the gym with his team?"

"He's probably here to spy on us," Alex whispered back as he and Cory got in line for their turn to do Skills.

"But he didn't know we would be here now. Don't you think it's strange that he wouldn't be with his team at a competition?"

"Well, not *that* strange," responded Alex, knowing how miserable it was to be around Nora and Kyle.

"Next!" shouted the field operator.

Alex and Cory walked to the side of the Skills field with Frodo. It seemed like Cory had practiced this routine thousands of times, but doing it for real was always nerve-wracking.

The Hobbits decided to do their driver runs first. Alex watched anxiously as each cube was collected and Cory moved Frodo around the field, putting cubes in towers. So far this run was going perfectly, and all that remained was for Cory to deliver the remaining stack of cubes into the scoring zone. If these cubes fell, there would be no time to try to restack them.

As Frodo approached the scoring zone, some of the students in the room began to cheer them on.

"Almost there!" called out a girl from the back of the room.

"You got this!" came a guy's voice from a short distance away.

Was that Euggie? wondered Cory, but then she refocused her attention back to the field.

As she completed the stack successfully and backed the robot away from the corner, several of the students started clapping for them.

"Good job!" said the field operator as he entered the score on his iPad and his helpers started resetting the field. "Since there aren't any other teams here waiting, do you want to go ahead and do another run?"

"That score is the best we can do with driving," replied Alex, "so I think we'll do our autonomous run next."

Cory nodded her head in agreement as she picked up Frodo and moved the robot to the starting position for their Autonomous Skills routine.

"Let me know when you are ready to go," said the field operator, once the field-setters had stepped out of the field.

After double-checking their robot, Cory looked up and said, "OK, I think we're-"

"Hang on," interrupted Euggie, jumping up and walking out onto the field. "This cube is in the wrong position." He reached down and moved a cube to the opposing side of a tower before returning to where he had been sitting on the floor.

"Thanks," said Cory, a bit surprised.

"Autonomous Skills," began the field operator, "in 3, 2, 1."

Normally when Autonomous Skills started, Frodo would immediately take off, going through the preprogrammed route around the field. But not this time. The robot remained still.

Stooping down to look at the robot again, Cory immediately recognized the problem.

"Frodo's battery is dead!" she said, looking up at Alex.

"Can we swap it out with our spare battery?" asked Alex.

"That *was* our spare battery," confessed Cory. "I forgot to plug in our other battery after our last match."

Alex took in a deep breath and let it out slowly, not knowing what to do.

"Hey, I think I have a battery you can use," Euggie called out. He reached into his backpack and, after feeling around for a moment, pulled out a battery. "Here ya go," he said, offering it to Cory.

"Thanks! You are a lifesaver today!" she exclaimed.

"No problem. Glad I could help," said Euggie.

Within moments, Cory had the battery changed and Frodo had begun its Autonomous Skills run.

As Alex watched the robot move around the field, he wondered how a nice guy like Euggie could be part of a team like The Best. Did Euggie have any idea that his teammates were cheating?

The autonomous run ended and, combined with the score earned in Cory's driver run, the Hobbits ended up with a very high score, probably enough to walk away with the title of Robot Skills Champion for Tumultuous Towers.

"Thanks for helping us out," Alex said to Euggie, as Cory pulled the robot off the field and returned the borrowed battery. "So how is your team doing?"

"Oh, I don't know," laughed Euggie. "I'll catch up with them at some point. Nora and Kyle like to do their own thing at competitions. I just build and journal. They handle everything at tournaments."

"Besides," Euggie added with a slight grin, "Nora can be a bit over-the-top at competitions."

Alex wasn't for sure yet, but he was becoming more and more convinced that Euggie had no idea what Nora and Kyle were up to.

Suddenly the door to the Skills room opened and in walked Nora. Right behind her was Kyle, carrying Destroyer. Apparently they were now ready to do their Skills runs.

Not wanting to engage with Nora and Kyle any more than they had to, Alex and Cory immediately made their way towards the exit. Nora gave them a cold, focused stare the entire way.

Once outside the Skills room, Alex looked at his watch. "We need to get back to the gym for our next match," he announced. And with that, the Hobbits took off for the queuing tables. They would have to discuss all that had happened once they had time to stop and catch their breath.

Chapter 11

Much to Alex's relief, none of the matches on the schedule put his team in an alliance with Nora and Kyle. Unfortunately, the Hobbits had to compete against The Best in their next match. Alex and Cory would be on the blue alliance, along with Heavy Metal. Cory noticed that their alliance partner's robot was quick and powerful, and it looked like they were utilizing an inertial sensor. Both Alex and Cory feared Heavy Metal would be a likely target of Nora, being much more of a threat than the Hobbits were themselves.

Ian, the driver for Heavy Metal, took the lead in putting together their strategy. "We'll focus on stacking as many cubes as we can in the scoring zones. You all put as many cubes in the towers as you can."

"How many can you all stack?" Cory asked.

"We can usually do at least two stacks of four or five," replied Ian, the driver for Heavy Metal.

"Sounds great!" said Alex, much more enthusiastically than he felt inside. How could their alliance even compete when their opponent had a kill switch aimed at their robots and was itching to use it?

As the Hobbits and Heavy Metal walked to the field and began putting their robots in their starting positions, Ian did a double take at his team's inertial sensor. The light on the port was flashing red. Heavy Metal huddled around their robot, trying to figure out what the problem was. But it was no use. There was no time to find a solution. The match was about to start.

"Red alliance, are you ready?" asked the MC.

"What's going on?" asked Alex. The red alliance on the other side of the field all gave a thumbs up.

"Not sure," said Ian. "Our inertial sensor isn't working."

"Blue team, are you ready?"

"We disconnected it for this match," continued Ian as he gave the thumbs up signal. "Our autonomous will be messed up, but we should be OK for driver control."

"Autonomous in 3, 2, 1."

The fanfare blared out the speakers, signaling the beginning of the autonomous period.

All four robots began their preprogrammed routines. But with the malfunctioning inertial sensor, Heavy Metal's robot ended up moving around the field in random directions, knocking cubes around rather than collecting them. Without a substantial fight from the blue alliance, the autonomous period ended with the six point bonus being awarded to the red alliance.

"We can turn things around now," said Ian. "We don't use our inertial sensor in driver control."

One can only hope, thought Alex.

"Driver control in 3, 2, 1."

Suddenly things begin to improve for the blue alliance. Heavy Metal moved quickly, plowing through cubes on the field, collecting almost every cube they came in contact with. They had delivered one large stack into the rectangular scoring zone and were well on their way to delivering a second stack.

Alex found himself starting to watch Nora more closely than the field. Surely she would hit their freeze button at any moment to prevent Heavy Metal from delivering this second stack. Looking up and locking eyes with Alex for a brief moment with an evil grin, Nora pressed the freeze button on her controller. But to her surprise, nothing happened. She pressed the button again, several times, but nothing happened. The color drained from Nora's face as she realized that their plan wasn't working. Alex was puzzled. *Why couldn't she freeze their robot?*

Although the scores between the two alliances ended up being very close for the driver controlled portion of the match, the autonomous win was enough to tip the match in favor of the red alliance. Despite his alliance's loss, it dawned on Alex that maybe The Best was not as invincible as they seemed.

Chapter 12

An integral component of VEX Robotics competitions is alliance selection. While alliances are randomly assigned during the qualification matches, teams get to select their alliance partner for the elimination matches. During alliance selection, all teams are lined up in order of their ranking from the qualification matches, with the highest ranking teams picking first. The invited team has the option of accepting or declining (in hopes of forming a more preferred alliance with another team). Much strategy is involved in alliance selection, and many teams spend time scouting in advance to determine which teams would make good partners. It is important to choose a team with a robot that both complements your team's strengths and offsets your team's weaknesses.

* * *

Alex and Cory finished their qualification matches ranked #14 and were chosen by the team ranked #9 during alliance selection. In their first elimination match, the Hobbits and their alliance partner played well, but ended up losing. On the bright side, Alex and Cory now had some free time to catch up on all that had happened while the rest of the tournament continued. With Frodo safely locked up in the wooden crate in their pit, they grabbed lunch from the concessions stand and made their way to the stands to watch the remaining matches.

"You know," said Alex as he sat down on the bleachers, "I find it very interesting that Nora's kill switch didn't work earlier when we were allied with Heavy Metal. I wonder what went wrong."

"I don't know," said Cory, "but I have a hard time believing that Euggie is involved in Nora and Kyle's cheating."

"I agree, especially after he helped us so much today. I honestly don't think he even knows about it. So what do you think we should do now?"

"I'm working on it. But, in the meantime, we need to gather as much information as we can. How about you video the rest of The Best's elimination matches? In fact, it looks like they are queuing up now, so you might want to go get a good seat. And while you are doing that, I'm going to have a chat with Euggie."

Cory was definitely taking charge.

As is typical for a competition, things never really slow down enough to enjoy lunch.

* * *

"So, how do you stand working with Nora and Kyle?" said Cory, rather directly.

Euggie turned to see who had joined him. "Oh, hey... Cory, isn't it?"

Cory nodded her head.

"Does it look like I'm working with them?"

"No, it looks like you're avoiding them."

"I'm not avoiding them," Euggie replied with a smirk. "At tournaments I just let them do their own thing."

Not satisfied with these evasive answers, Cory pressed on. "Why? Don't you enjoy competing?"

"Of course I do."

"But, you're content to build and journal - and then sit back while Nora and Kyle run the show?"

Euggie didn't respond.

"You're obviously really good at building," continued Cory, "because Destroyer is solid and always performs well. But I would think it's aggravating to let Kyle and Nora take over on competition days."

"Oh, you have no idea."

Now she was getting somewhere. Cory waited for Euggie to elaborate.

"They think they are better than everyone else, and they like to say so every chance they get. If they ever lose a match, they blame it on anyone and anything but themselves."

It was like the dam had finally broken and Euggie was suddenly releasing all of his frustration.

"I spend time building the robot, and then they confiscate it and program it without me. But, when there's a mechanical problem, they always have to come back and get me to fix things. It's such a pain to work with them," Euggie admitted sadly, "that I'm actually thinking about quitting robotics altogether next year."

"How would you like to put Nora and Kyle in their place?" offered Cory. "If we work together, we could turn things around."

"How would we do that?" asked Euggie, becoming intrigued with where this conversation was going.

"I'll call you once I get the details worked out," said Cory. "What's your phone number?"

Euggie gave her his number. At that moment, Cory had a brilliant idea.

"Euggie! Do you think you could-"

Cory leaned over and whispered the rest of her request in his ear.

Euggie hesitated for a moment before nodding his head.

"Hey, can we join in on this too?" came a voice from behind them.

Cory turned around to see two young boys who looked practically identical except for the fact that one of them was wearing glasses. Apparently they had overheard the conversation.

"Sure," said Cory slowly. "What are your names?"

"I'm Luis," replied the one wearing glasses, "and this is my brother Loren. We're the Sonic Boomers."

"Well, Luis and Loren," said Cory, "Glad to have you on board."

"We don't know what we can do to help, so just let us know. But for now, we've got to get to our next match. We're in the final!"

After giving Cory their phone numbers, Luis and Loren left for the queuing tables.

Euggie sighed. "I guess I should get down there with my team too. After all, it *is* the final."

"OK, we'll talk later. Hopefully you can complete that little favor I asked. Then we'll be able to make some progress!"

Chapter 13

Rejoining Alex who was videoing the last semifinal match, Cory could see a robot frozen on the field, yet another victim of Nora and Kyle's cheating. The Best and their allied team, Heavy Metal, were in the process of securing another win, which would take them to the final match.

Cory wanted to tell Alex about the new recruits she was assembling to take on The Best, but the crowd was getting louder in anticipation of the final, making it more difficult to hear. "I just met the Sonic Boomers," shouted Cory, trying to talk over all the noise. "Luis is the one on the far side of the queuing table next to the team's robot, or is that Loren?" she said, squinting her eyes.

"And their alliance partner is the Robonauts," Alex yelled back. "I did a bit of research on them. Corbin is their programmer, and he's working on some neat stuff with the vision sensor. Kenzie and Emma seem to do well with the journaling and building."

Then it dawned on Alex. *It's the sensors!* The disconnected inertial sensor was most likely what prevented Nora from being able to freeze Heavy Metal's robot earlier in the day. "The Robonauts need to unplug their vision sensor!"

"Why?" asked Cory. "What good would that do?"

"I think Kyle and Nora are targeting sensors! Remember when I overheard them talking in the lab? I heard them mention something about exploiting a vision sensor. I bet that unplugging it would keep Nora from being able to freeze their robot."

"Seriously?" exclaimed Cory. "You need to get down there and tell them! Right now!"

At that moment, the lights on the competition field brightened. The MC picked up the microphone to announce the teams who would be competing in the final match.

"There's no time," said Alex. The final was about to start.

But as the ref was preparing to begin the match, Kyle made the time-out signal - initiating a three minute delay.

"That's just what we need!" Cory said excitedly. "Now, go! The clock is ticking!"

Prompted by a firm push from his enthusiastic teammate, Alex hurried out of the stands. This would be awkward. *What am I supposed to say? "Hi, good to meet you, The Best is about to freeze your robot, but you can avoid this by disconnecting your vision sensor."* They would probably just think he was crazy. And how could he talk to the Robonauts without Nora and Kyle getting suspicious?

As Alex was pushing his way through the crowd, he saw Kenzie suddenly leave the competition field and run off toward the gym exit - with Emma yelling something about getting rubber bands. Reaching into his pocket, Alex confirmed that he still had a number of extra rubber bands. Maybe this was the break he needed? Moving as quickly as he could, he caught up to Kenzie near the gym door.

"Need some of these?" Alex offered Kenzie the rubber bands he had in his pocket. She took them eagerly. This was good! But when Alex tried to explain about the vision sensor, Kenzie was so rushed that she was only able to offer a quick "thanks" and "gotta go" before rushing back to the field. This was bad. But at least he had tried to warn her.

"How did it go?" asked Cory, as Alex returned to his seat.

"Not good. I tried to warn her, but there wasn't enough time."

Only a few minutes passed before the Robonauts' robot froze on the field, just as Alex and Cory had feared it would. Knowing that it would happen in advance didn't make it any easier. They painstakingly watched as the ref declared The Best's alliance as the Tournament Champion of Tumultuous Towers.

Chapter 14

Now that the final match was over, the crowds settled down and everyone's focus moved to awards.

"I sure hope your plan for dealing with The Best is good," whispered Alex.

"Me too," replied Cory. "But you will be happy to know that Phase One is already in motion."

"Phase One?"

"Phase One is recruitment. We have the Sonic Boomers who want to join our cause. And we have Euggie helping us. Being on Nora and Kyle's team, he can get close to them." Cory raised her eyebrows for emphasis. "But we still need an advanced programmer."

"I bet the Robonauts would want to help, especially if they discover that the title of Tournament Champion was stolen from them today. And Corbin, their programmer, is *really* good."

Alex paused as the head ref announced, "The Robot Skills award is given to a team who has the highest combined score of driver and programming skills. This award also qualifies the winning team for the State Championship. And the Robot Skills award goes to the Hobbits from Cedar Ridge Academy!"

Alex and Cory walked to the front where their hands were promptly shaken and they were handed a trophy. Turning around, several photographers (who always seem to appear out of nowhere during an awards ceremony) took their picture. After the last camera had flashed, the Hobbits returned to their seats.

The next trophy presented was the Design award, which was given to Heavy Metal. The Hobbits clapped and cheered for them. It was wonderful to see a team win honestly and on their own merit. When the trophies for Tournament Champion were awarded, however, Alex and Cory were more reluctant to join in the applause, since they knew The Best did not deserve to win. Nevertheless, Nora soaked up the attention, making herself prominent in every picture taken. For Nora, this was what it was all about. As hard as this was to watch, it only made the Hobbits more determined than ever to bring The Best to justice.

"The final trophy is the Excellence award. The Excellence award is the highest award presented at a VEX Robotics competition." The head ref went on to describe all the criteria that had to be met to receive this award.

"And the Excellence award goes to... The Best!"

As the crowd once again broke into a loud applause, Nora, Kyle, and Euggie stood up and began filing out of the bleachers one by one, passing right by Alex and Cory. As Euggie passed by Cory, he lightly tapped her on the shoulder and discretely passed her something. She took it and passed it to Alex.

"What's this?" asked Alex, looking at the small blue flash drive in his hand.

"Phase Two," answered Cory. "Thanks to Euggie, we have a copy of the code that Nora and Kyle are using to freeze robots. Now, we need to figure out how it works. So your next assignment is to find a way to contact Corbin from the Robonauts and see if you can get him to help us with the code."

Little did they know how valuable the flash drive in Alex's hand would be.

Chapter 15

In real life, it is impossible to freeze a robot through hacking as Kyle does in this story. And it is certainly impossible to fry a robot's brain by zapping it with battery power through the field system. So, VEX teams, don't get any ideas. One of the goals of every team should be to conquer problems through innovation and creativity while always playing fair and following the rules.

* * *

Valley Creek High robotics lab
A week and a half after Tumultuous Towers

Slurrrrp.

"Maybe we need to up the torque on the motors." said Corbin, setting his Dr. Pepper down.

"Don't even think about it. Over there," instructed Emma, as she pointed outside the field without looking away from the robot. She was not about to have sticky soda accidentally spilled all over the field.

Corbin sighed and moved his drink.

"The last time I tried a 100rpm cartridge, it didn't make a bit of difference," Emma continued, studying the intakes on the robot.

"It brought in the cubes slower and still hung up on the fourth cube."

"There's gotta be a way to pick up six cubes. Still no replies on the forum," Corbin said, as he glanced over at his laptop sitting on the floor. He had posted the problem on the VEX Forum last night and was hoping another team could offer some advice.

"I don't know," said Emma, as she started the motors and watched Columbia collect the cubes - again, hanging on the fourth cube. "I'm starting to think that stacking more cubes is impossible."

Since they were out of ideas for dealing with The Best's cheating, the Robonauts had let it go for now. Skyscraper, their next tournament, was coming in less than a week, and there was much work to do, particularly in getting Columbia to stack more cubes.

Emma had built a longer tray that could hold up to six cubes. This tray folded over cleanly so it would fit within the initial 18" x 18" x 18" size limit and pop into place at the start of a match. But now they were stuck on being able to intake the six cubes needed to fill the tray.

"What's impossible?" asked Kenzie, walking into the robotics lab and plopping her books on the table.

"Still trying to get this silly robot to pick up more than four cubes."

"Have you tried increasing the torque on the motors?"

"You know what torque is?" joked Emma.

"Yes, believe it or not, even I, the mere keeper of the journal, know what torque is," Kenzie responded dramatically. She laughed, "Just don't ask me to explain it."

"More torque isn't the answer." said Corbin.

"What if you tried lowering the angle of the tray?" Kenzie suggested. "That way the motor wouldn't have to work as hard."

"But if we lower the tray angle very much," said Emma, "it's going to stick out a mile behind us, and I'll end up whacking everything on the field when I turn around."

"Well, there's got to be a way to do this," said Kenzie, making her way over to the field and nudging Corbin out of the way. Getting down on her hands and knees in front of the robot, she asked, “Can you try running it again? Maybe we’ll see something we’ve missed so far.”

Corbin collected the cubes they had been using to test, and Emma repositioned the robot. Just as before, Columbia had no trouble until the fourth cube, where it was unable to bring any more in.

“Can you leave the motors running for a minute?” requested Kenzie.

She studied the intakes. The flaps on the tread were rubbing against the cubes, but they just weren’t pulling them up.

“Maybe the intakes need to be at a different angle?” Kenzie pressed them to be more parallel to the tray.

As soon as she did this, the fourth cube flew up the tray, followed by the fifth and sixth cubes.

All three of the Robonauts broke into huge grins.

“Glad this journaler could be of service!” teased Kenzie.

“Yeah, whatever,” laughed Corbin. “Let’s run it again.”

As Emma and Kenzie lined up the cubes in front of the robot, Corbin explained, “I don’t think it has to do with the angle of the intakes, but with the fact that Kenzie was pushing them in. It just needed a bit more-”

“Compression!” Emma blurted out. “I bet we need to add more rubber bands!”

Sure enough, another run with additional rubber bands brought all six cubes into Columbia’s tray.

“Well, I’m glad this *programmer* could be of service!”

Emma rolled her eyes. “It was a team effort.”

“I’ll definitely include this breakthrough in the journal,” remarked Kenzie. “Who would have thought a couple of rubber bands could solve such a big problem?”

“Which is why one of us needs to make sure we *always* have *plenty* of extras at tournaments,” Emma said emphatically. “That way we won’t have to send Kenzie on another crazy last-minute mission right before a match.”

“Yeah, no kidding.” agreed Kenzie.

Corbin glanced at his watch. “We only have a few minutes left.”

Kenzie looked up at the clock on the wall. There was over thirty minutes until school would let out. "What happens in a few minutes?"

"Didn't you hear?" asked Emma. "There's a pep rally in the gym today, and Coach Warren asked us to help."

"Help *how* exactly?" asked Kenzie.

"Check this out," said Corbin, quickly removing all the cubes from Columbia's tray. "Can you hand me that flag over there?"

Kenzie looked over and saw a school flag lying on the table. She picked it up and handed it to Corbin, who promptly positioned it on the very top of the tray on the robot.

Emma nudged Corbin with her elbow. "I guess some of my building skills are finally rubbing off on you, huh?"

Corbin shrugged nonchalantly.

"Also," added Emma, "Coach Warren thought it would be fun to invite the middle school team to the pep rally. They're bringing their robot too."

Just then, Coach Warren stuck his head in the lab. "Are you all ready?"

"Yes." Emma picked the robot up off the field, and the Robonauts began gathering their things.

As Corbin was closing his laptop, he noticed that he had received a new direct message on the forum. It was from someone with the username Alex_Hobbits. The message gave a phone number and said, *Call me soon. Important. I'll explain everything.*

What that's about? wondered Corbin. But it would have to wait for now. The Robonauts needed to get to the pep rally.

Chapter 16

The Robonauts followed Coach Warren out of the lab. Waiting at the end of the hall was the visiting robotics team from Valley Creek Middle School. The three boys were standing in a line, dressed in their team's competition-day uniforms - green army hats and camouflage. With them, was their robot, a small modified clawbot with tank treads. Their coach, Mrs. Myers stood next to them.

"Team Alpha," Coach Warren said, "it's good to have you with us again."

"Yes, sir! Good to be here, sir!" said the team leader, stepping forward and giving a respectful salute.

"At ease, Jake," laughed Mrs. Myers. "It's good to see you all again."

"Here, Jake," said Coach Warren, handing him two small school flags. "See if you can find a spot for these on your robot."

Jake immediately handed the flags to his teammates, who knelt down and wedged them in the robot so that they pointed out either side.

"Aren't they cute," whispered Emma to Kenzie as the entire group made their way to the gym. "Those flags make the robot look like it has ears."

"OK, here's the plan," instructed Coach Warren. "Once all of the members of the basketball team are announced and the cheerleaders move to the front, drive your robots back and forth in front of them. Got it?"

The Robonauts gave a thumbs up, and Team Alpha gave a resounding, "Yes, sir!" Pulling a small two-way radio out of his pocket, Jake ordered, "Alpha 2, put the robot in position to deploy."

Kenzie smiled and rolled her eyes as she heard Jake's voice echo out of the paired radio attached to the hip of Alpha 2 standing no more than three feet away.

Alpha 2 set the robot on the floor and readied his controller. "Copy that, Alpha 1. Robot is now in position," he replied over the radio, again from no more than a few feet away.

Music began playing out the speakers, and a loud voice sounded, "Let's hear it for the Valley Creek basketball team. Go Vikings!" As the names of each member were announced, the team came running onto the court, dribbling and showing off for the crowd.

On cue, the cheerleading squad began doing cartwheels and flips across the gym.

"Here we go," said Emma, as she powered on the controller and moved the joysticks forward. Columbia zoomed out onto the gym floor in front of the cheerleaders.

"Alpha 2, deploy!" Jake said, again over the radio. Immediately, Team Alpha's robot took off behind Columbia. The two robots blazed across the gym floor with the school flags waving in the air.

After several minutes of cheering and total mayhem, the pep rally began to wind down.

"Come out to CRA on Friday night to support our basketball team as they go head to head against the Cedar Ridge Lightning Bolts. And let's hear it for our robotics team who will be competing this Saturday!"

The Robonauts stepped forward and waved to the crowd.

"Did he say *The Lightning Bolts*?" asked Kenzie. "Isn't that the home school of The Best?"

"It sure is," Emma confirmed.

Corbin turned and offered each member of Team Alpha a fist bump. "Thanks for helping! We'll see you at Skyscraper this weekend."

Skyscraper would be held on Saturday at a community college four hours away and in a different time zone. The combination of the long drive and the time change was just enough to make the tournament an overnight trip. The plan was to meet at the school and head out shortly thereafter early on Saturday morning. Very early. After the long day of competing, the team would then stay the night at a local hotel before returning home the next morning.

As the gym was clearing out, Kenzie returned her focus to the upcoming robotics tournament. Looking directly at Emma and Corbin, she announced, "It's time for a team meeting."

Best working with us undercover. We're calling ourselves *The Resistance.*

Corbin raised his eyebrows to ask the obvious question. *Do we want in on this*?

With emphatic nods from both Kenzie and Emma, Corbin replied, "We're in. Can you send us The Best's code?"

"Yes, I'm sending it to you now."

Corbin picked up his phone and found a new message from Alex with a file attached.

"Got it."

"Great, keep us updated," Alex said, ending the call.

"Well," said Corbin, "we have lots of work to do and not much time to do it. So... y'all up for pizza?"

Leave it to Corbin to put his stomach at the top of their to-do list - and right in the middle of the afternoon, nonetheless. But some hot pizza *would* be perfect right about now. And Pietro's Pizza *was* right across the road.

"Sounds good to me," said Emma.

"Me too," said Kenzie.

"Sweet, let me make a call to postpone my ride, and I'll run Columbia back to the lab. You all go ahead and order the usual. I'll meet you there in 10 minutes."

Chapter 18

Pietro's Pizza always had fast service. Before Corbin could join the girls, the waitress had already brought the pizza and a large pitcher of Dr. Pepper to the table. Soon after, Corbin walked in and sat down on the opposite side of the booth from Kenzie and Emma. He pulled his laptop out of his backpack and opened it up on the table.

"It may take me some time to look through this code," said Corbin, as he grabbed his first slice of pizza.

"Whatever it takes," Kenzie said. She was preoccupied with her own project of completing her math homework for the day.

Emma nudged Kenzie with her elbow. "Look who just came in."

"Who?" asked Kenzie, raising her head out of her math book.

Emma pointed toward the entrance to the restaurant. Gordon had walked in with a couple of friends.

Maybe he won't see us, thought Kenzie. She continued to work on her math, with Corbin occasionally pecking the keys on his laptop and Emma texting on her phone. Suddenly a voice interrupted Kenzie's train of thought.

"Hey, Kenzie."

She looked up to find Gordon standing beside their table.

"Um... hi," Kenzie said, reluctantly.

"How's the math homework going? I hear you might need some help with it?"

"How did you-" Kenzie started to ask. She stopped when she saw the smile on Emma's face. *So that's who Emma was texting.*

"Oh, I'm actually finished now," said Kenzie as she closed her book.

"But she needs you to look over the last problem," Emma inserted as sincerely as she could. "It was pretty hard."

"I'm glad to!" said Gordon.

Kenzie gave Emma a good kick under the table as they both scooted down a bit to make room for their new visitor to sit with them.

While all this was going on, Corbin remained absorbed in his laptop and was oblivious to the conversation.

Kenzie reopened her math book and pulled out her papers. "Here's what I have."

"Let's see," said Gordon, as he began reading the problem out loud. "Martin is turning part of his backyard into a chicken pen."

Kenzie planted her face in her hands as Gordon continued.

"His backyard is a rectangle that measures 20 feet by 40 feet. He wants to put a chicken wire fence that stretches diagonally from one corner to the opposite corner. How many feet of fencing will Martin need?"

Gordon took a minute to look over Kenzie's work and the answer she had come up with. "Well, 2000 feet of fencing is much more than is needed. You need to take the square root of 2000, like this..." Gordon proceeded to show her how to finish up the last step of the problem to arrive at the correct answer. "So, Martin will need approximately 44.72 feet of fencing for his chickens."

"Wonderful," Kenzie replied. "The next time I need to build a chicken fence in by back yard, I'll know just what to do."

Emma put her hands in front of her mouth to hide her smile.

"Great!" said Gordon. "Let me know if you ever need any more help with math. Here's my number."

Gordon picked up Kenzie's pencil and jotted his phone number on one of her papers. With that, he got up and returned to his table.

At that moment, Corbin looked up from his laptop and blurted out, "I know how The Best is freezing robots!"

"How?" asked Emma.

"Their code captures network packets that flow through the field system, and then it backtraces the embedded information to determine which robot to target."

Emma wrinkled her nose. "In English?"

"Basically, Kyle found a bug in the field control software, and he's used it so Nora can freeze robots by pushing a button on the controller.

"Can we prevent Columbia from being frozen by disconnecting our vision sensor like Alex suggested?" asked Kenzie.

"Maybe as a one-time fix," said Corbin. "It does look like The Best was targeting certain robots at Tumultuous Towers based on a specific sensor. But I wouldn't count on them doing this in the future. I'm sure they'll change their code to exploit any sensors being used, including the encoders which are built into the V5 motors. That means all robots on the field with The Best will be vulnerable.

"Have you found a way to sabotage their code so that The Best freeze their own robot?" Emma asked.

"Yes. All we need to do is hardcode the ID number of their robot's brain in the program," replied Corbin.

"So we need to somehow get their brain ID," jumped in Kenzie, who was starting to follow the new plan.

"That's what I'm thinking," Corbin confirmed.

"Then, let's get it," said Kenzie, "and soon. We leave for Skyscraper in..." she paused to do some quick calculating, "just under 84 hours!"

"See, you *can* do math," Emma remarked.

"Only when it's pertinent."

"Well, this is definitely pertinent," Emma agreed, "and we don't have much time, so we need to get moving."

Corbin leaned back in his seat. "But it's not like we can just waltz into their lab, find the brain ID, and reprogram their robot."

The conversation fell silent. The Robonauts were at a loss of what to do.

After a moment, Kenzie snapped her fingers. "Yes, we can."

"And how exactly would we do that?" asked Corbin.

"We can go to CRA on Friday night for the basketball game. And, it would be understandable if we happened to *accidentally* get lost in the halls while looking for the restrooms... and end up in The Best's robotics lab."

With an huge smile, Emma gave Corbin a salute. "Alpha 1, we have a plan."

"Copy that," replied Corbin, returning the salute. Now, we need to work out this operation."

Over the next hour and with a refill of Dr. Pepper, the Robonauts put together their plan to infiltrate the robotics lab at CRA. On Friday night, they would take action.

Chapter 19

Friday evening, the night before Skyscraper

The rest of the week went by slowly for the Robonauts, but the day of the basketball game at CRA finally arrived. The school van was carrying all of the students from Valley Creek High who had decided to go. Among the students on this van were the Robonauts, filling up the back row. Gordon was sitting one row up from the Robonauts, disappointed he hadn't been able to sit with Kenzie.

Forming a team huddle in the back seat, Corbin reminded Kenzie and Emma of the plan. "First, we have to locate The Best's robotics lab. Then we have to find their robot, get its brain ID from the menu, plant the ID into their code, and download the changes."

"How long is all that going to take?" asked Emma.

"Only a few minutes," answered Corbin, "but you two are going to have to keep watch while I'm in the lab."

It was about a 45-minute drive to Cedar Ridge Academy. The first thing that Kenzie noticed when they arrived was that CRA was much larger than Valley Creek. The building was two stories high with multiple wings.

As everyone got out of the van and began making their way towards the gym, Kenzie whispered to Emma and Corbin, "How are we ever going to find the robotics lab in this humongous building? I bet we'll have to search this place room by room."

"Or maybe not. Look up there," Emma whispered back as she pointed to a room up on the second level. "Whaddya bet that's the room we're looking for?"

Hanging in a window on the second floor was a banner that proudly displayed *Home of The Best*.

* * *

The buzzer sounded, and the first quarter of the basketball game began. The Valley Creek Vikings started off well, taking control of the ball after the tip-off.

The Robonauts were sitting near the exit so they could slip out as inconspicuously as possible.

Looking around the gym, Kenzie saw lots of people, several that she recognized from robotics tournaments. She was kind of hoping to run into Alex, but he was nowhere to be found. He was probably home being responsible and getting some rest before tomorrow's tournament. Kenzie thought it was sweet that he had gone out of his way to try to help her team at Tumultuous Towers. It would be good to thank him in person. She found herself disappointed that Alex wasn't there.

"I didn't know you liked basketball," said Gordon, trying to start a conversation with Kenzie.

Gordon had finagled a seat between her and Emma, with Corbin sitting behind them on the next row.

"Yeah, I do," Kenzie replied.

"Interesting. But I thought robotics was your thing?" Gordon continued.

"Oh, robotics *is* my thing," replied Kenzie. "But robotics and basketball are a lot alike, you know."

"Really? How so?"

"They are both competitive team sports. Both require an enormous amount of preparation and practice. Both have referees. And both are fun. But you learn a ton more in robotics."

Gordon had never thought of robotics as a sport before.

Now that the game was under way, it was time for Corbin to make his move. He slowly slipped his backpack over one shoulder. Once both teams were battling on the far side of the court and everyone's attention was turned away, Corbin left the stands and slipped out of the gym where he would wait for Kenzie. The plan was for her to slip out separately to assist him in finding and infiltrating the lab.

"So, how much time do you spend preparing for a robotics tournament?" asked Gordon.

Kenzie could talk about robotics all day, but there were more pressing matters at hand. How was she going to be able to break away from the conversation? Making eye contact with Emma, Kenzie silently pleaded, *Could you please assist here?*

"Hey Gordon," Emma interrupted the conversation, "Do you think you could help me with a math formula to determine some distance measurements for a path our robot needs to travel." She pulled up a picture on her phone of the team's robot sitting on their practice field. "We need Columbia to go from here to here," she continued, using her finger to point out a path.

"Sure," said Gordon, not wondering at all why someone would want to discuss math at a basketball game.

As Gordon and Emma hovered over the phone discussing math, Kenzie slipped out of the gym unnoticed, joining Corbin in the foyer.

"Sorry it took so long," said Kenzie. "I had trouble breaking away."

With that, the two Robonauts rushed off to find The Best's robotics lab.

Chapter 20

The halls at CRA were dark other than a few emergency lights that stayed on all the time. Using the flashlights on their phones to see, Kenzie and Corbin walked quickly, taking the first flight of stairs they came across. As they emerged from the stairwell and onto the second floor, they found themselves in a wing of classrooms with long hallways on either side of them.

Kenzie shone her flashlight in each direction. "Which way do we go?"

"I don't know," said Corbin. "But we'll probably find the lab faster if we split up. You go that way and I'll go this way."

Taking off in opposite directions, they began checking each room they came to, looking for The Best's robotics lab.

* * *

"And based on the value you get from this equation," explained Gordon, "you can figure out the exact distance your robot needs to travel. And in order to find this angle, you can use the tangent function..."

Emma needed to text her team to find out how things were going. But how could she with Gordon sitting right there going on about trigonometry?

"By the way," said Gordon, changing the subject, "did you see where Kenzie went?"

Emma shook her head.

At that moment, one of the Valley Creek Vikings on the court fell to the ground. Or rather, was *pushed* to the ground.

"Foul!" yelled the referee, who proceeded to award the Vikings a free throw.

The crowd fell silent as the foul shot was launched. The ball bounced off the backboard and straight down through the net.

"Yes!" yelled Emma, throwing one fist up in the air.

As the game continued, Emma suddenly realized that the seat next to her was now empty. *Uh-oh. Where is Gordon?* In all the excitement, she had lost track of him. Looking around, she caught a glimpse of Gordon walking out of the gym. Emma immediately sent a group text to Kenzie and Corbin.

How's it going? Kenzie, I think Gordon is looking for you.

* * *

The first several rooms that Kenzie checked appeared to be regular classrooms. But in the next room, her phone light revealed a scattering of orange, green, and purple. *Tower Takeover cubes!*

"Hey! I found it!" she whispered as loudly as she dared down the hall toward Corbin.

The light of his phone immediately changed direction as he doubled back to join Kenzie.

"After you," said Corbin.

"Gee, thanks," replied Kenzie as she entered the room and flipped on the light switch. There was a full-sized robotics field in the center of the lab. Around the edge of the room were tables, chairs, and large storage cabinets. One of the storage cabinets was prominently labeled *The Shire*. Directly across the room from The Shire was another cabinet, adorned with a skull and crossbones.

"That's gotta be it," said Kenzie. Opening the cabinet door with the skull and crossbones, she removed a robot, the same one that she remembered from Tumultuous Towers, with a DR4B lift. Kenzie placed it on a table next to the cabinet. "Yep, this is definitely The Best's robot. I'll head back downstairs and stand guard while you work."

"Text me if anything goes wrong," said Corbin, as he pulled his laptop out of his backpack and powered up the robot.

Leaving the lab, Kenzie made her way down to the base of the stairs on the first floor to keep watch.

Suddenly, "Eye of the Tiger" began to blast. *Ugh*! She had forgotten to silence her phone. After stopping the music as quickly as possible, she looked at her incoming text. It was a warning from Emma. Gordon was coming.

* * *

Gordon was standing outside the doors of the gym looking for Kenzie. *Where did she go?* He walked over to the door leading to the parking lot and looked out, but she wasn't there. And the only sign of life in the foyer was the guy manning the concessions stand. *Maybe he saw her come through?*

"What can I get for you?"

"Popcorn, please." *Maybe I can share it with Kenzie when I find her*, thought Gordon as he fumbled through his pockets for quarters to pay. "Have you seen a girl come through here in the last few minutes? She has long, brown hair."

"Yeah, she came through. I think she went that way," he replied, pointing to the hallway on his left.

"Thanks," said Gordon, taking the popcorn and heading toward the hall where he had been directed. The hallway was dark except for a faint light a short way down.

Just as he was about to give up and head back to the gym, he heard music suddenly blaring from down the hall, but it stopped almost as soon as it had started. *Was that "Eye of the Tiger"?*

"Kenzie?" Gordon called, as he started walking slowly down the hallway.

* * *

Kenzie froze as she heard her name. *How am I going to explain this to Gordon?*

As he continued to approach, it became obvious he was not about to turn around. Kenzie proactively came out from the edge of the stairway and into the hall to meet him.

"You know, the restrooms are really hard to find in this place," fibbed Kenzie.

"Aren't there restrooms next to the concession stand?"

"Oh, I must have missed those."

Gordon looked down and shuffled his feet. "So, would you like to go for a walk? It's really nice outside."

Could his timing be any worse? thought Kenzie. But what was she supposed to do?

"Uh, sure. I guess."

As they walked down the hall, Kenzie texted Emma and Corbin, *Hallway clear at the moment, stuck with Gordon.*

Going well here, adding the brain ID to the code now, replied Corbin.

Gordon and Kenzie made their way outside and began walking down the long sidewalk that led around the perimeter of the parking lot.

"Would you like some popcorn?" Gordon offered. "And if your chilly, you're welcome to wear my jacket."

Apparently I'm going to have to set things straight with him, thought Kenzie. *But how can I tell him to back off without hurting his feelings?*

"Gordon, we need to talk," Kenzie started. "I want you to know that-"

Kenzie felt her phone vibrate. "Hang on a second."

It was a text from Emma. It read, *Everything OK? Having fun with Gordon?* followed by a string of laughing emojis.

Ignoring Emma's message, Kenzie tried again. "I want you to know that..."

She paused. Out of the corner of her eye she saw a small red sports car pull into a parking spot near the entrance to the building. Nora was driving. And Kyle was with her.

What if they're on their way to the robotics lab? thought Kenzie.

"...that I really like walking. Let's go faster." *Maybe I can catch up to them before they make it to the building.*

"Sure," said Gordon, a bit confused. As they took off, Gordon accidentally dropped his popcorn. *Oh well,* he thought as he glanced back at it, *I just need to keep pushing forward.* Kenzie was becoming difficult to keep up with.

As they sprinted across the parking lot, Kenzie came to the realization that Nora and Kyle were too far away and that she was not going to be able to reach them before they entered the building.

"Hold on, Gordon." said Kenzie as she suddenly stopped. "I gotta send a quick text."

Gordon obliged, grateful for the chance to catch his breath.

Kenzie hurriedly sent another text to Emma and Corbin. *Nora and Kyle are here! Emma, get out here and stall them! We can't let them catch Corbin in the lab!*

Chapter 21

Emma rushed outside and spotted Nora and Kyle.

"Hey!" she yelled, walking up to them.

They both stopped, looking at Emma as if to say, *why are you talking to us?*

"I remember y'all from Tumultuous Towers! I wanted to tell you that..."

Say something good, Emma thought to herself.

"...that I think you are doing an outstanding job in robotics this year. You are really living up to your team's name." Emma cringed as she forced out the words.

"Well, our talent comes naturally," said Nora, eating up the compliment.

"Aren't you one of the Robonauts?" asked Kyle.

"Yes, I am. So, how did you all get to be so good at robotics?"

"Oh, I pick up on things quickly, and I'm just better than other people. It does help matters that I really know what I'm doing..." Nora stood there going on and on about all of her glorious virtues.

I can't keep them here forever, Emma thought.

Kyle, who didn't like to hear Nora go on about herself any more than Emma did - especially since he was the true brains of the team - was getting rather impatient.

Corbin needed to hurry.

* * *

Kenzie was sprinting toward the building's entrance where Emma was stalling Nora and Kyle, with Gordon doing his best to keep up.

"Look," Kenzie said, pointing ahead, "there's Emma and some of my friends from robotics. Let's stop and see what they're up to."

"Oh, hi you two!" Emma called out to Kenzie and Gordon as they ran up to the group.

"Hey Emma!" said Kenzie, as she stopped. "Just getting in some cardio."

Gordon, a bit out of shape, was bent over with his hands on this knees, breathing heavy and recovering from his impromptu run.

"Nora was just telling us that you have to excel in lots of areas to be good in robotics," said Emma, hoping to keep Nora talking.

"Yes, that's right," continued Nora, "like physics, programming, math-"

"Math?" interrupted Gordon. He looked up, still out of breath, but as always, eager to discuss math. "Oh... I love math... especially trigonometry... Do you all... use any trig functions... with your robot?"

Nora had no idea what a trig function even was. She paused, not knowing how to answer.

"We need to get inside," Kyle said, taking the opportunity to move things along. "We're already late for the game."

Kenzie and Emma glanced at each other in relief. Nora and Kyle were not headed to the robotics lab.

"OK, see you all tomorrow morning at the tournament, bright and early," said Emma.

As Nora and Kyle opened the door to enter the building, Kyle happened to glance up to the second floor of the school. "Is that a light on in our lab?"

"It looks like it," said Nora. "I wonder who could be up there this time of night."

"Probably just the cleaning crew," replied Kyle, "but let's stop by and check it out before we go to the gym."

Upon hearing this conversation, Kenzie and Emma immediately alerted Corbin.

Nora and Kyle are headed to the lab! texted Kenzie.

Get out of there now!!! texted Emma.

Gordon, now breathing easier from his sprint across the parking lot, just stood there. He still had no clue what was going on and was thinking, *These girls sure do text a lot*.

* * *

Corbin had just finished modifying The Best's code when his phone buzzed. It was the warning texts from his teammates. Quickly, he compiled and downloaded to the brain. *Done. Now time to get out of here.*

Corbin hurried over to the door and peeked out into the hallway. Flashlight beams from two phones were emerging up the staircase. There would be no time for him to get out of the lab without being seen. Frantically looking for a place to hide, he ran over to The Shire and yanked the door open. Half of the cabinet was empty, and there was just enough room for him to squeeze inside, but there was no room for his backpack. Corbin glanced around the room looking for a quick place to stash it - and across the room he saw an open window. *That'll work!* He dashed to the window and flung the backpack outside so that it dropped to the ground. Then he scurried back to The Shire and squeezed himself inside with his laptop, pulling the door shut.

A split-second later, the door to the lab opened and Nora and Kyle strolled in. The first thing Kyle noticed was that Destroyer was out of the cabinet and sitting on the table.

"What's the robot doing out of the cabinet?" asked Kyle.

Nora glanced around the room, but didn't see anything else out of place. "Euggie must have left it out."

Kyle picked up Destroyer and placed it back in the cabinet with the skull and crossbones.

"So how is our secret weapon coming along?" asked Nora.

Listening from inside the cabinet, Corbin's heart raced.

"Oh, it's done." replied Kyle. "I just need to enable it in our code when we're ready to use it."

"So how does it work?"

"Well, we no longer have to find a specific sensor ahead of time to exploit. Since all robots have motors, I targeted the built-in encoders."

I knew it! thought Corbin from inside the cabinet.

"Nothing will change for you," continued Kyle, "except that you will also have the ability to fry our opponent's robots with buttons X and Y."

"Frying our opponent's robot..." mused Nora with her evil grin. "No one will have a chance against us."

"But you can't use it every match," said Kyle.

"And why not?"

"Because people will get suspicious if every robot we go up against is suddenly damaged. And, if you fry every opposing robot, there wouldn't be any left to compete with in the elimination matches. We need to use it in only the most dire of circumstances."

Although Nora liked the idea of literally destroying the competition, Kyle was probably right. Besides, they should be able to beat all the teams at Skyscraper without frying robots. After all, they were *The Best*.

"Fine," said Nora reluctantly.

Corbin sat there listening to all of this from inside The Shire. After another minute or so, the lights went out in the room, and Corbin heard the door close. Climbing out of the cabinet, Corbin texted his teammates, *Mission accomplished. Can you all pick up my backpack for me? It's outside below the robotics lab.*

* * *

Later that evening, on the way back to Valley Creek

"Did everyone enjoy the game?" asked Coach Warren from the front of the van.

None of the students responded. Their enthusiasm would have been much higher if their team had won. Still, the Valley Creek Vikings had played very well. Too bad the Robonauts didn't get to see much of the game. By the time they made it back to the gym after their mission in the robotics lab, there was less than a minute left on the clock. The score ended up being close with Valley Creek losing by four points.

As for Gordon, he had finally stopped pursuing Kenzie, at least for tonight - much to her relief.

With the Robonauts huddled again in the back seat of the van, Corbin relayed everything that had happened in The Best's robotics lab to his teammates.

"You hid in a cabinet called the Shire?" Emma laughed. "That's too funny!"

"Was Frodo there?" asked Kenzie. "Wasn't that the name of the Hobbit's robot?"

"Yeah, Frodo was there. It wasn't the most pleasant experience," said Corbin.

"We ought to start calling you Bilbo Baggins," Emma joked.

Changing the subject quickly in hopes that his new nickname would not stick, Corbin spent the next few minutes describing the secret weapon that he had overheard Nora and Kyle discussing. Kenzie and Emma listened intently as Corbin described how they planned to fry any robot that got in their way of winning. The very thought of robots being damaged like this was disconcerting to say the least. But this was what they were up against.

As the Robonauts were contemplating their situation, it dawned on Emma. "Do you all realize that Skyscraper is our last tournament this fall? At the moment we aren't signed up for any tournaments in the spring. So, if we are going to qualify for the state championship, Skyscraper is our last chance.

Kenzie smiled. This was the perfect time to share her news.

"Oh, we'll have one more chance after Skyscraper," she said.

Emma and Corbin stared at Kenzie.

"How?" asked Corbin.

Kenzie leaned in and lowered her voice. "You know the guy we saw Coach Warren talking to at Tumultuous Towers?"

"You mean, Curly?" asked Emma.

"Yes. It turns out his name is Jeffrey. He is our regional rep from REC Foundation."

"I like the name Curly better," Emma commented. "I vote we stick with Curly."

"OK, Curly it is. Anyway, I overheard Coach Warren and Curly talking. They are planning a new tournament for the spring. But you won't believe the best part - the tournament's going to be hosted at Valley Creek!"

"Oh, how fun!" Emma exclaimed.

* * *

It was just after 10:30pm when the school van pulled back into the Valley Creek parking lot. Most students would head home to crash for the rest of the weekend. But not the Robonauts. They had to be back at the school in less than 6 hours to load the van and leave for Skyscraper. Such was the life of a robotics team.

As Kenzie and Emma began the short walk to their homes, they received one final text from Corbin.

<Corbin>
Thanks for watching my back and covering for me tonight.

<Kenzie>
Any time, my friend.

<Emma>
Thx for all you did tonight. Good night, Bilbo.

* * *

When Kenzie made it home, she was worn out. It had been a difficult day. For that matter, it had been a difficult week. Still, despite the late nights, the hard work, and all the problems with The Best, Kenzie loved everything about robotics. Being a part of the Robonauts was worth it. She reached over to the alarm clock by her bed and set it to wake her up in plenty of time to get ready and make it to the school by 4:00am. But before she went to sleep, she picked up her phone. She wanted to send a quick text to someone who had been on her mind.

Chapter 22

Earlier that same Friday Evening, the night before Skyscraper

Dribbling forward, Euggie leapt into the air and pushed the basketball into the goal, completing a perfect layup. Then, moving further back, he stopped, aimed, and launched the ball towards the goal, swishing it through the net.

"Awesome!" Cory yelled, walking into the CRA gym with Alex.

"Here, you try," said Euggie, bouncing the ball over to Cory.

"Nah," said Cory, catching the ball. "I don't do basketball. It's not my thing."

"Ah, sure you can!" Euggie said with a smile.

"Give it a try," encouraged Alex.

Reluctantly, Cory carried the ball onto the court and threw it toward the goal the best she could. The ball bounced off the backboard to the left, and Euggie ran up to catch the rebound.

"See, I told you I wasn't any good at this."

"I'm sure you would get better if you played more often," said Euggie. "Alex, do you want a shot?"

"Nah, I just wanted to return this." He pulled out the small blue flash drive containing Kyle's code and handed it to Euggie. "There wasn't much we could do with it, but we've passed the code along to someone who can help."

Euggie glanced back and forth to be certain Nora and Kyle were not around. "Great, I'm glad it helped." He dribbled a couple of times and made another shot. "So, are you two coming to the game tonight?"

"We plan to," said Cory, "but we have lots of journaling to finish first. We'll eventually make it because we have to get everything packed up for the tournament tomorrow."

"When is your team packing?" asked Alex.

"Oh, Nora and Kyle will probably pack early tomorrow morning right before we leave."

"They don't let you help?" asked Alex.

"Nope," Euggie said as he shot the ball again. "Other than building the robot and keeping the journal, the two things they can't do well for themselves, they don't let me do much of anything."

"That is just wrong," exclaimed Cory. "Do they even let you drive the robot?"

Euggie shook his head no.

"Alright, stop by the robotics lab tonight and you're gonna try driving. I'll show you a few things and help you get started."

Euggie stopped and looked at Cory.

"Um... Nora and Kyle would be furious! And besides, I don't think I'd be any good."

"I'm sure you would be better if you drove more often," said Cory, using Euggie's line from earlier.

"Point taken." Euggie started shooting again. "Sure, I'll stop by after the game. But we can't stay long cause it's gonna be an early morning."

* * *

It was the middle of the fourth quarter when the Hobbits finally arrived at the game. In an effort to avoid Nora and Kyle whom they saw sitting in the bleachers, they found seats on the far side of the gym and began to cheer for Euggie's team. After the buzzer sounded that signaled the end of the game, Euggie ran over to the sideline to grab a water bottle.

"Great job, Euggie," said Cory. "You played well!"

"Thanks! Give me 10 minutes and I'll meet you two up in the lab."

He then turned around and headed off to the locker room.

Looking around, Cory noticed the group from Valley Creek who were making their way out of the gym.

"Isn't that the Robonauts?" asked Cory.

"Yeah, it sure is," said Alex. "Let's go see them." He wanted to find out how things were going with the code he had sent. And, truth be told, he wouldn't mind saying hi to Kenzie.

"Hold on," Cory cautioned. "Nora and Kyle are right over there." She motioned to the left with her head. "I don't think it would be a good idea for either of us to be seen talking to them right now."

Cory was probably right. After all, he *had* just passed Kyle's code to the Robonauts a few days ago in hopes that they could figure out how to sabotage it. The less connected they appeared to be to the Robonauts, the less chance there was of Kyle and Nora figuring out that they were working with them.

"OK, then let's just head up to the lab," said Alex, watching Kenzie leave the gym out of the corner of his eye.

* * *

Upstairs in the robotics lab, Cory had rolled out the wooden crate that they used to transport Frodo. She was packing it with spare parts, tools, the journal, a power strip, and a couple of extra batteries. While she was putting all of this together, Alex set Frodo on the field. He always liked to confirm that their robot was still working well the night before a tournament.

"So, do you think Corbin is making any progress with the code you sent him?" asked Cory.

"I sure hope so. If not, Skyscraper may end up being a repeat of Tumultuous Towers, with robots being frozen left and right." Alex powered up Frodo and began driving the robot around the field, pushing cubes and picking them up.

"You know," said Cory, "since we'll have Destroyer out on the field tonight, we ought to have a one-on-one match with Frodo. A fair match with no freezing. Could be fun."

"Are you serious?" said Alex.

"Well, Euggie *is* on their team. He has every right to drive his own robot around."

"Just be careful," said Alex. "If Nora and Kyle ever found out, they would be furious."

"Oh, it's not that big of a deal," said Cory, moving over to the cabinet with the skull and crossbones. As she opened the cabinet and reached in to pick up the robot, the door to the lab opened. A booming voice from behind her said, "Hey! What are you doing touching my robot?"

Cory's heart jumped as she froze in place. Turning around, she saw Euggie standing there, doubled over laughing.

"Come on Euggie! That was so mean!" She walked over and gave him a playful shove.

Alex was laughing too. "I told you to be careful."

"OK, Euggie, let's try this again." She picked up Destroyer, and brought it to the side of the field opposite from where Alex was working.

Powering up the robot, Cory browsed through the different programs available on the controller.

"Here we go, let's try the program called *Competition Drive*. That sounds like it will work."

Cory started the program and began pushing the joysticks in different directions to see how it would make the robot move. Once she figured out the pattern, she showed Euggie how to control the robot.

"Now you try it."

Euggie took the controller. At first he moved the robot slowly and a bit erratically, but he gradually got the hang of it and began to drive it smoothly.

"Just don't press any buttons," said Alex. "You never know what might happen."

"Hey Alex, I'm coming after Frodo," Euggie said playfully.

Euggie aimed Destroyer at Frodo and began zooming toward it.

"No way are you gonna catch me," said Alex, as he countered the moves and took Frodo in the opposite direction.

The two robots circled the field several times in a high speed chase, weaving in and out of towers and plowing through cubes. It wasn't exactly the match that Cory had envisioned, but at least Euggie was having fun.

"See," said Cory, "I knew you could drive!"

"I haven't caught him yet," said Euggie.

Destroyer drove Frodo into a corner and then proceeded to barrel forward toward the robot at full speed.

At the last second, Frodo moved out of the way and Destroyer crashed hard into the corner of the field, bouncing off the wall and flipping over on its side. The hit caused the field perimeter to jolt all the way around.

"OK, you win," said Euggie.

"I'm not sure which is worse," Cory said with a laugh, "Nora's freeze button or your driving!"

"Freeze button?" asked Euggie, setting Destroyer upright.

Before Cory could answer, Euggie spotted a problem on Destroyer. The right intake was bent.

"Uh-oh. I'm going to pretend I don't see that," said Euggie.

"See what?" asked Alex.

"Check out this intake."

Alex and Cory sat down with Euggie to examine the injured robot and assess the extent of the damage. The robot would still bring cubes into the tray. And the lift would still raise cubes up in the air. But when they attempted to stack, the cubes would hang on the right intake.

"Yikes!" said Euggie, "that ain't good."

"Think you can fix it?" asked Cory.

"Probably, but not by tomorrow."

There was silence for a moment while they thought about this development.

"You know, with all the trouble Nora and Kyle have caused, it kinda serves them right," said Cory.

"Yeah," said Euggie, "but they are my team. As much of a pain as they are, I really didn't mean to cause this.

Another pause.

"But," continued Euggie, "it's not like we can do anything about it at this point. And it *will* be fun to watch Nora's face tomorrow."

This was something all three could agree on.

"Now," said Euggie, "what's this about a freeze button?"

Euggie was amazed to hear about the code that Kyle had written, allowing Nora to freeze a robot in its tracks. Although it didn't really surprise him that his teammates would stoop to this level.

"It's getting late," Cory finally said with a yawn, "and we need to call it a night. We have to be back here in just a few hours."

"Is everything packed up?" asked Alex.

"Yes, everything except what's out on the field," replied Cory. "Let's go ahead and pack up Frodo."

With that, Alex and Cory put Frodo in his box along with the controller so that everything was ready to go.

Euggie picked up Destroyer and put it back in the cabinet on the left side of the shelf along with the controller, ready for Nora and Kyle to pack up in the morning.

After a quick straightening up of the room and the field, all three exited the lab. They went downstairs and to the parking lot to wait for their rides. Euggie's parents were already there and Cory's parents pulled in only a few moments later, leaving Alex waiting by himself.

Alex sat down on the curb, enjoying the nighttime air and admiring the moon, shining clearly without a cloud in the sky. As he sat there, he felt his phone buzz.

<Kenzie>
Hey Alex! It's Kenzie. Whatcha up to?

<Alex>
Sitting here waiting to be picked up. Saw you tonight at the game.

<Kenzie>
You were there??

<Alex>
Not until late in the 4th quarter. Wish I could have made it over to see you.

Kenzie paused for a moment before replying.

<Kenzie>
I would have liked that.

<Alex>
Me too.

<Alex>
Cory was afraid that Nora and Kyle would think we were plotting with you all against them.

<Kenzie>
LOL we are. But I understand.

<Alex>
Any luck with the code?

<Kenzie>
Yes. It's a long story. Make sure you are watching when Nora and Kyle start their first match tomorrow.

<Alex>
I definitely will.

Alex and Kenzie continued to text for another 10 minutes or so until Alex's parents pulled in the parking lot.

<Alex>
Looks like my parents are here.

<Kenzie>
K, time for me to crash anyway.

<Alex>
See ya at Skyscraper.

After receiving a final smiley face and robot emoji, Alex hopped in the car and headed home to get a few hours of sleep.

Chapter 23

4:00am Saturday morning at CRA

Alex and Cory pulled the wooden crate with Frodo in it out from under the table.

"Weren't we just here?" said Alex, still trying to wake himself up.

"Shh..." Cory said softly with a finger over her lips when she saw who was coming into the room.

The lab door opened and in came Nora and Kyle. Kyle went right to the storage cabinet, opened it, and started gathering everything they needed to take with them to the tournament. Nora stood there, letting Kyle do all the work.

Cory walked to the lab door and held it open for Alex as he rolled their crate out into the hallway. As the door was swinging shut, they both heard the first words of the day from Kyle.

"Hey, Destroyer is on the wrong side of the cabinet."

Alex and Cory glanced at each other, but then continued down the hallway, moving a bit faster. Since they had this huge crate with them, they had to use the elevator at the end of the hallway to take it downstairs. They passed the stairway just as Euggie was coming up the steps, headed to the lab.

"You don't want to go that way," said Cory.

"Why?" asked Euggie.

"This way," she replied, taking his arm and pulling him along with them down the hall.

All three entered the elevator and Cory waited until the doors closed.

"Did you put Destroyer back in the cabinet last night, just like you found it and in the same spot?"

"Yeah, I think so," said Euggie, not understanding what Cory was getting at.

"Is it possible you put it back on a different side of the cabinet?"

"Umm... Possibly."

"Well, Kyle just found the robot in a different spot and now they know that someone moved it."

"Oops," said Euggie, not seeming to appreciate the severity of the situation.

"Aren't you concerned that Nora and Kyle are gonna start asking questions?"

"Yeah, I guess. But honestly, I'm over dealing with Nora and Kyle. And I've decided to quit the team after today's tournament."

"What?" asked Cory and Alex at the same time.

"I've decided to quit," repeated Euggie.

After the elevator doors opened, Alex pushed the crate out into the hallway and they moved towards the doors that led to the parking lot.

"Let's just get through today for now. We'll talk about this later," said Cory, realizing that it was not the time or place for this discussion. Deep down, Cory knew that she couldn't let Euggie quit. Not only was Euggie playing a crucial role as part of The Resistance, but the truth was, he loved robotics. And he was good at it. It would be a shame and a waste for him to quit because of Nora and Kyle.

Chapter 24

The trip to Skyscraper was a good opportunity for the Robonauts to get a few extra hours of sleep along the way. With only one quick bathroom stop along the way, the Robonauts made good time, arriving at their destination and pulling into the parking lot just after 7:30am. They wouldn't be able to check in to their hotel until after the tournament was over. Fortunately, the hotel was only a few minutes down the road. Being the only hotel nearby, this is where all the robotics teams would likely stay after traveling long distances for Skyscraper.

Skyscraper was the largest tournament that the Robonauts had ever seen. The tournament was set up in a huge three court gym area. There were three raised competition fields in the center of the room, several practice fields available, and multiple large screens for displaying tournament stats throughout the day. There was a small camera mounted high up on a tripod over each field to capture and livestream each match. Students from various teams were all over the room, coming and going from the pit area. The Robonauts saw many teams they recognized, including Heavy Metal, the Sonic Boomers, Team Alpha, and of course, the teams from CRA.

After getting their pit set up, the Robonauts had a few minutes before the tournament would start and just enough time for a team picture. They found the perfect location for this at the side of the gym where a Skyscraper backdrop was hung.

"Corbin, why don't you kneel down in the center with Columbia?" asked Emma.

"And then we can stand behind you on either side," added Kenzie.

Never one to argue with the girls on things like this, Corbin knelt down and the team posed while Coach Warren took their picture.

"Now," said Coach Warren, winking at the girls, "do something crazy."

Corbin complied, making the silliest face he could for the camera. Emma and Kenzie, who had planned ahead for this very moment, discretely whipped out a small poster with the words "Team Geek" in large letters. They held this up over Corbin's head while pointing their thumbs inward at him.

Once the picture had been taken, the girls hid the poster as quickly as it had been produced. As they walked back to their pit, Kenzie gave Emma a fist bump. Corbin never had a clue.

According to the schedule, qualification match #4 was the first match for The Best. The Robonauts would be free to watch since they didn't have their first qualification match until #9. As Kenzie went through the schedule, highlighting all the matches for her team, she paused at #56. The Robonauts and the Hobbits were on an alliance. And they would compete head to head against The Best on the opposing alliance.

* * *

As qualification match #3 came to an end, the Robonauts were well positioned in the bleachers to watch what was coming next. When Alex and Cory arrived, they joined the Robonauts, sitting behind them in the bleachers.

"This is Alex, our rubber band guy," said Kenzie, officially introducing him to Emma and Corbin for the first time in person. "And you must be Cory."

"That's me!" said Cory.

"Good to finally meet you two," said Emma.

Without wanting to directly mention the elephant in the room, Alex worded his question carefully. "So, any progress with our little project?"

"Oh yeah," said Corbin emphatically.

Alex and Cory sat there waiting for Corbin to elaborate. But he didn't.

"Just watch this match," said Kenzie.

* * *

Nora, Kyle and Euggie stood beside the competition field for qualification match #4, their first match of the day. When the autonomous period began, Destroyer started going through the normal routine of collecting cubes and attempting to stack them within the 15 second time period. It successfully collected six cubes and then moved to the scoring zone to deliver them. But rather than a nice stack forming as it usually did, the cubes hung up on the bent intake.

"What on earth?" screeched Nora.

"What happened?" said Kyle, looking over at Euggie.

Euggie shrugged his shoulders and shook his head.

Within a few seconds, the autonomous bonus had been awarded to The Best's opposing alliance and the driver control portion had started.

"Something's not right with Destroyer," yelled Nora. "I can't stack cubes!"

Although The Best's alliance partner was putting up a good fight on their own, the opposing alliance continued to maintain a significant lead. Nora's only hope of winning would be to freeze the opposing robots and hope that her alliance partner could pull their team ahead before the end of the match. She immediately pressed the freeze buttons, targeting both of the opposing robots. But the robots continued moving as if nothing had happened.

"Uggh!" screamed Nora.

Kyle's jaw dropped, not knowing what was going on.

I guess I'll have to play defense to try to prevent them from scoring, thought Nora.

But, when she tried to do this, she made another startling discovery. Destroyer was completely unresponsive. Frozen.

Nora was mad. And Kyle was irritated. How could Destroyer have been frozen?

Unable to top the score of the opposing alliance, The Best lost the match.

As the crowd cheered for the winning alliance, Nora looked in the bleachers and noticed Alex and Cory sitting there. With the Robonauts. And they all looked happy about what had just happened. Alex and Nora briefly locked eyes and, with a surge of confidence and bravery, Alex mouthed the words to Nora across the room, "Not. The. Best."

Nora clenched her fist in fury.

"Yes!!!!" said Corbin, giving Emma a fist bump.

"Good job, Corbin," said Kenzie, giving him a high five over Emma's head.

Indeed, it appeared that Kyle had finally met his match. Corbin had out-programmed him. This time.

"I take it you all had something to do with this?" asked Cory.

Suddenly realizing they were being a bit too transparent, the Robonauts regained their composure.

"Might have," said Emma with a smile and a wink.

"Y'all, we gotta go," said Kenzie. "It's time for us to queue up for our next match."

As the Robonauts left the bleachers and made their way to the queuing tables with Columbia, Kenzie's phone buzzed.

<Alex>
You all ROCK!

<Kenzie>
Thx. But this day isn't over yet.

Chapter 25

"What on earth happened?" yelled Nora at Kyle.

The Best was in their pit, and Kyle was fumbling through the code on his laptop, trying to figure out how Destroyer could have been frozen. So far Kyle had no answers.

"And, why can't Destroyer stack cubes?" yelled Nora at Euggie.

"It looks like this right intake is bent," said Euggie, making a good show out of studying the robot closely.

This was definitely not Nora's day. Nora was used to everything going her way. And if things were not going her way, she would *make* them go her way. But not today. She had completely lost control of things.

"Euggie, did you work with the robot last night?" asked Kyle. "I noticed it was in a different spot this morning when I pulled it from the cabinet."

"No," replied Euggie, as he continued to work with the right intake, "no building last night." Which was absolutely true. Now, driving and crashing the robot was an entirely different story.

"Well, *something* has happened to the robot since earlier this week," continued Kyle. "And code doesn't just stop working. I'm gonna download my code to the brain again and see what happens. Euggie, keep trying to get that intake working 'cause our next match is coming up soon against the Robonauts and the Hobbits."

Nora perked up. "Against the Robonauts? Euggie, do you know anything about that team?"

"I've seen them around. I think this right intake is now working a little bit better," continued Euggie, trying to change the subject. "Give it a try?"

Nora picked up the controller to test out the intakes. They were now working somewhat better. Not perfect, but probably good enough to deliver a stack successfully, even if the stack ended up being shorter or a bit wobbly.

"Good," said Nora.

Still thinking about the Robonauts and the Hobbits whom they would face in the upcoming match, Nora turned to Kyle. "I think we need to deploy our secret weapon in our next match."

"Agreed," said Kyle, as he scrolled to the pertinent section of code, enabled it, and then downloaded to Destroyer's brain.

"Secret weapon?" asked Euggie, "What's that?"

"Oh, you'll find out," said Nora.

"Great," said Euggie, "I can't wait to see it. Whatever it is."

As Kyle continued to type on his laptop with Nora looking over his shoulder, Euggie decided it was time to slip out. "Time for a bathroom break," said Euggie as he left the room. Once outside in the hall, he texted Cory as quickly as he could.

* * *

"So tell us about your robot."

The judge had an official name tag hanging around his neck and a clip board in his hands.

"Our robot uses a design that we both worked on," said Cory. "The drive train system is geared for speed, but we had to make sure that the motors don't overheat during a match." As Cory was talking, her phone buzzed in her pocket. With her right hand she reached down to silence it and, without missing a beat, continued on for a moment or so before passing the conversation over to Alex. Alex promptly began explaining the intake system, how it worked, and how it came together. After a few more follow-up questions, the interview was over. Writing down a few notes, the judge said "thank you," and moved on to find other teams that he needed to interview.

Looking at her phone, Cory saw that it was a message from Euggie. Her eyes widened as she read his text.

"We've gotta get this message to the Robonauts!"

"Why, what's wrong?" asked Alex.

"It's The Best. They're pulling more of their stunts and fighting back."

Looking around the gym, they didn't see Kenzie or Emma or Corbin anywhere.

Pulling out his phone, Alex sent a quick text to Kenzie. *Where are you?*

Sitting on the floor by the wall, across from the queuing tables, replied Kenzie.

"Follow me," said Alex, as he grabbed Frodo and took off. Cory followed, both of them moving as quickly as they could.

Finding the Robonauts, Alex and Cory sat down with them in a circle.

"Hey alliance partners!" said Emma, "Ready to discuss our strategy?"

"Hang on," interrupted Cory. "We have a problem. We just heard from Euggie and he says that Kyle has re-downloaded his code to Destroyer."

"Ugh, I was afraid he might do that," said Corbin.

"What does that mean?" asked Kenzie.

"It means that the changes I made the other night are no longer part of Destroyer's programming - and Nora will now be able to freeze robots again."

"But that's not the worst of it," said Cory. "Euggie says that Nora is really mad and that she is planning to use their secret weapon in this match."

"Oh no," said Corbin, thinking back to the night in the cabinet when he heard firsthand about their secret weapon. "That's the last thing we need right now. I need my laptop from the pit ASAP."

"On it," said Kenzie, immediately jumping up and running off to get it.

"I'll go with you," said Alex, taking off right behind her.

As Kenzie and Alex arrived at the Robonauts' pit, a judge came up to them. This judge was a woman, tall and thin, with glasses and her hair in a bun, and a clipboard in her hand.

"Hello, I'm Tiffany! Are you from team..." She paused to check her clipboard. "Robonauts?"

"Yes, ma'am," said Kenzie. While she was always glad to see a judge, this was terrible timing.

"My two teammates are queuing for a match right now. Would you like me to get them?"

"No, that won't be necessary."

As Kenzie continued listening to the judge - and without breaking eye contact, she picked up Corbin's backpack from the table and handed it to Alex. Taking the backpack, Alex slowly moved away. Once he was far enough away from the important conversation that had just ensued, he sprinted back with Corbin's laptop.

Dropping to his knees next to Corbin, he set down the backpack. Corbin promptly pulled out his laptop and started typing. "How are we doing on time?" asked Corbin.

"Only a couple of minutes until we need to check-in at the queuing table," said Cory. "Nora and Kyle are already there."

"Where's Kenzie?" asked Emma, looking around for her.

"Still in the pit. She got caught by a judge."

Once the updated program had been downloaded to Columbia's brain, Emma and Corbin moved to the queuing table, along with Alex and Cory.

"Nothing like last minute changes," said Emma.

On the opposite side of the queuing table, Nora and Kyle both had stern faces expressing their readiness to dominate in the upcoming match. Euggie was with them, although he was much more calm. In fact, he seemed disconnected from it all. As usual, Nora and Kyle were running the show. The Best was allied with a team that the Robonauts had never seen before. The team was named Monstrosity and consisted of two guys and one girl. Their robot was named Leviathan, and printed on their team shirts was its definition: a*nything of immense size and power.*

While they were waiting for the field to be reset from the previous match, Emma whispered to Corbin, "So, what did you change in the code?"

"You know how Kyle's code targets a robot based on its unique brain ID? Well, I enabled a bit of code I wrote earlier this week just in case we needed it. It's basically a loop that changes the brain ID every second. Hopefully this will make Columbia immune to Destroyer's attacks."

"Corbin, you are a genius!" said Emma, patting him on the back.

"I just wish I could do something to protect the Hobbit's robot," Corbin lamented.

Emma glanced over at Frodo. It would be devastating if Alex and Cory's robot was damaged.

At that moment the MC announced, "And now for qualification match #56. We have The Best and Monstrosity on the red alliance and the Robonauts and the Hobbits on the blue alliance." As the two alliances walked to the competition field with their robots, Kenzie finally returned and ran up beside them. Slowing down and falling in pace with Emma and Corbin, Kenzie took a quick second to catch her breath before sharing the news.

"We're getting a second interview!"

Chapter 26

The ref looked over the field. All the cubes were in place. The scorers with their iPads were standing in their positions. Everyone was wearing their protective goggles. After confirming that both alliances were ready, the MC announced, "Autonomous in 3, 2, 1." The buzzer sounded, indicating the start of autonomous.

As all four robots popped out from the expansion limit to take on their full size and begin their preprogrammed routines, it became obvious why Monstrosity had named their robot Leviathan. The tray on their robot was enormous. It swung up, almost whacking the cubes positioned on the field in front of it. This tray had four stages, looking like it could hold ten to twelve cubes.

The Robonauts were successful in stacking three cubes and the Hobbits were successful in pushing one cube into their scoring zone, giving the blue alliance a total of four points.

Leviathan picked up six cubes and delivered these in a perfect stack, just to the outside of the scoring zone. Obviously they still had work to do on their autonomous programming.

The Best had collected five cubes and Destroyer was on its way to the scoring zone to deliver these. The stack formed nicely until it came time to stack the fifth cube, which hung up on the right intake. Destroyer backed up from the stack it had formed with the fifth cube still in its tray right as the buzzer sounded that ended the autonomous portion.

Although the result was obvious, the ref still walked to each side of the field to examine the cubes in the scoring zones. Then the ref walked back to his spot by the side of the field and made the sign for "tie". When two teams tied during the autonomous part of the match, the protocol was to split the points awarded to the two alliances.

Without any further delay, the MC continued, "Driver control in 3, 2, 1."

All the robots started moving again on the field.

"Ready to use our secret weapon?" asked Nora, as she moved Destroyer around the field.

"Let's do it," said Kyle.

Nora pressed the Y button on her controller to fry Columbia. When she did this, Columbia jumped slightly but then kept moving as if nothing had happened. With each successive press of the button, the same thing happened again. The effects were negligible.

The relief on Corbin's face was evident. "My code worked! It looks like we're gonna be OK." Corbin just wished that he had made the new loop in his code repeat every half-second instead of every second. Maybe this would have prevented the robot from jerking so much? But this could be fixed before their next match.

"We can still win this," said Kyle. "Especially if Monstrosity is able to score some large stacks."

Monstrosity already had twelve cubes collected in its tray and it was still within the first 30 seconds of driver control.

But one of the problems with a large tray of cubes is that it makes the robot very top-heavy. If a team is not careful, an overloaded tray can cause their robot to tip over, especially if no counter-tipping measures are in place. And this appeared to be the case with Leviathan.

The robot spun around toward the scoring zone but never completed the turn. Rather, halfway through the turn, the robot began to sway sideways - and then crashed, falling on its side. Cubes went flying across the field and one landed right beside

Nora. Even worse, Leviathan's long tray was now lying horizontally across the field, completely blocking the rectangular scoring zone for the red alliance.

Nora couldn't believe it. She wasn't able to fry the Robonauts' robot, Destroyer wasn't stacking like it should, and her alliance partner was proving to be completely worthless to her. As the anger welled up inside of her, she let out a loud, "Hmmmph!" and threw her controller down to the ground. Then she left the field, stomping away to who knows where.

Kyle and Euggie looked at each other, at first not knowing what to do. But then Euggie picked up the controller and started driving Destroyer.

Meanwhile, the girl from Monstrosity, still in a bit of shock over Nora's temper tantrum, yelled down to The Best, "Sorry about that!"

Euggie turned his head toward her and, with a smile, said, "Oh, it's OK. It could happen to anyone. We'll make the best out of it." After all, that *was* their team name.

Kyle stood there with his hands on his hips, not even believing how all of this was playing out.

Still inexperienced with driving, Euggie fumbled a bit at first while trying to gain control of Destroyer. In the process, he inadvertently hit button X. Kyle had programmed button X to target the robot positioned on the left opposing side of the field, which in this case was Frodo. A completely vulnerable Frodo. Frodo got zapped right as it was in the process of speeding across the field. It caused the robot to stop responding and left Cory unable to steer or brake. Totally out of control, Frodo ran smack dab into a tower and spun out around it, knocking several cubes away that were lying in the vicinity. Without realizing exactly what had happened, Cory kept pushing the joysticks on her controller. But Frodo wouldn't move.

"Alex, Frodo isn't responding," said Cory in a panic.

"Try re-starting the controller," said Alex.

But a quick restart of the controller did nothing.

Looking over at the Hobbits and knowing full well what had just happened, Kenzie's heart sank. She watched them as they stood there helpless. There was nothing that could be done.

With 60 seconds still left in the match, the final battle between the two alliances would now be between Destroyer and Columbia, with Euggie and Emma driving. After reaching the painful conclusion that Frodo was not going to move again, Cory and Alex gave up and watched the rest of the field to see what would happen next. Cory was unsure whom to root for. On one hand, it *was* Euggie driving. And The Best *was* the Hobbit's sister team. But on other hand, the Robonauts had done a spectacular job today of dealing with Nora and Kyle. With all Euggie had been through this season, Cory finally decided that she hoped The Best's alliance would win. For Alex, the choice was much easier. He hoped that Kenzie's alliance would win.

Since Euggie was not experienced in collecting and stacking cubes, he decided that the most productive path forward was to prevent Columbia from scoring. He did his best to position Destroyer in Columbia's way and to block the blue alliance's scoring zone.

As the match continued, something unexpected and wonderful was happening. For the first time this season, playing against The Best had become fun for the Robonauts! There was no cheating or threats or intimidation. Despite the difficult start to the match, both teams were now giving everything they had to try to win. This high level of competition was exhilarating. And Euggie was having the time of his life. Only a few hours ago, Euggie was ready to completely quit robotics. But now that he was experiencing just how much fun it could be, Euggie thought to himself, *How can I possibly quit?*

After spending most of driver control battling against Destroyer, Emma finally snatched up three cubes. With the game clock counting down and only 22 seconds left, Kenzie yelled to Emma, "If you are going to stack those cubes, you need to do it now!" Emma backed Columbia up several feet from Destroyer - who was blocking the scoring zone - and plowed forward as hard as she could, ramming into The Best's robot. But Euggie was able to keep Destroyer stationary so that Columbia could not get by. Emma kept ramming Destroyer to try to get through and, on the third hit, Columbia hit so hard that one of the cubes it was carrying tumbled out of the tray. Bouncing off Destroyer, the cube landed in the rectangular scoring zone right next to the cubes that were already there. This put the Robonauts ahead by one point. As soon as this cube came to rest, the buzzer went off signaling the end of the match.

Immediately the crowd burst out clapping and cheering. Kenzie and Emma jumped up and down with excitement. Alex and Cory, still in disbelief over what had happened to Frodo, did their best to celebrate that their alliance had won this difficult match. The match had indeed been contentious and very close. Walking to the other side of the field, Euggie smiled and gave each of the Robonauts a hug. He was genuinely happy for them. Even though his alliance lost, this match had still been more fun than he had ever known robotics to be. This is what it was all about.

"Guys," said Euggie directly to Alex and Cory, "I'm so sorry about what happened to your robot."

"Thanks," said Cory, "But it wasn't your fault."

It was true that it wasn't directly Euggie's fault. But he still felt badly about it.

Noticing that Kyle had left the field with Destroyer and had headed in the other direction, Euggie decided to walk away with Alex, Cory, and the Robonauts. They made their way over to the bleachers and sat down.

"I want you all to keep doing your best," said Cory, holding back her tears, "and don't worry about us Hobbits. We'll be OK. But we need to put a stop to this."

"Yes we do," said Euggie softly. Alex and the Robonauts all nodded their heads in agreement.

"I need everyone's phone number," continued Cory, "so we can schedule a meeting and coordinate our efforts." Cory collected everyone's number and then said, "OK, back to the tournament! We'll be here cheering you on!"

With Frodo's brain damaged, the Hobbits were unable to continue competing. As a courtesy to their assigned alliances in the qualification matches, they still went through the motions of going to each match. But, other than this, they ended up being spectators for the rest of the tournament. At least the Hobbits had been able to complete their Skills runs earlier in the day. Alex and Cory would have much work to do in the coming weeks to repair Frodo. And they would definitely need Corbin's help to program Frodo so that it would be invisible to Destroyer. The only silver lining in all of this for the Hobbits was that they had already qualified for state.

Kenzie gave Cory a quick but warm hug. Then she and the Robonauts headed back to their pit.

But Euggie stayed put. He had a couple of additional things he wanted to say.

"First," said Euggie, as he moved right beside Cory, "thank you for all the encouragement you've given me." Indeed, Cory had been the one to encourage Euggie to do something about the situation with Nora and Kyle. She was also the one who had helped him discover that he was good at driving.

"Second, remember when I said I was gonna quit robotics after today? Well, I've changed my mind. All of this is simply too much fun to give up. So, whatever the plan is from this point forward, you can definitely count me in."

"That's good to hear!" said Cory, "cause we're gonna need your help in dealing with Nora and Kyle."

"Speaking of Nora and Kyle," said Euggie, "I should probably go find them."

"Good luck," said Alex, as Euggie took off to find the rest of his team.

Alex and Cory were now more determined than ever to put an end to Nora and Kyle's antics. As it turned out, both the CRA teams and the Sonic Boomers were also staying the night at the same hotel down the road. This would be the perfect night and location to meet. Tonight at the hotel would be the first official meeting of The Resistance.

* * *

Euggie found Nora and Kyle back at their pit. Kyle was staring at his laptop, trying to figure out what had happened. What had gone wrong and why hadn't Columbia been fried as planned? Not able to offer much assistance, Nora was just sitting there, stewing over all that had happened.

"Kyle tells me you took over and drove after I left," said Nora.

Uh-oh, thought Euggie, *I hope I'm not in trouble*.

"Yes, I did. I just wanted to make the best of things and try to help our team win."

There was a short pause.

"Thanks," said Nora, much to Euggie's relief.

"I'm also told that you fried Frodo during the match," continued Nora. "Good job. The Hobbits are not going to recover from that quickly." Euggie was definitely proving useful. "At least *someone* was doing something to help today."

Kyle glanced up from his laptop for a moment, but then decided not to respond. *Shouldn't he get some of the credit for designing their secret weapon in the first place?*

Despite the fact that he had stepped up to drive and had fried Frodo, Kyle still wasn't so sure about Euggie. How did he even know about their ability to fry robots in the first place?

"In our first match, Destroyer somehow froze itself," said Kyle, glancing at Euggie a bit suspiciously before continuing. "Since that first match, my code has worked fine against any robot we have chosen to target, except for Columbia. Somehow that robot seems immune to our attacks. Any idea why that might be?"

"No, but I'm sure you'll get it figured out," replied Euggie, growing more and more uncomfortable with Kyle's suspicion. At least he was on Nora's good side. For now.

Chapter 27

The Robonauts stood in the hallway outside the gym with their robot, waiting to enter the judges' room. It was almost time for their second interview. Waiting was definitely not something that Kenzie enjoyed. She would much rather jump in and tackle the interview head-on. She tried to think through all the questions the judges could ask and how her team should respond. Although the Robonauts had rehearsed a number of talking points, you never know what specific questions the judges might ask.

When the door finally opened, Tiffany, the same judge that Kenzie had talked to earlier, motioned for the team to enter the room. Inside the room, the Robonauts found two additional male judges sitting behind a long table covered with clipboards and papers.

"Welcome! Congratulations on receiving a second interview!" said Tiffany as she took a seat at the table.

"Thank you," said Kenzie. The Robonauts then proceeded to introduce themselves and explain the role that each member of their team had played in the process of designing, building, programming, and documenting their robot.

"So our first question," Tiffany began, "what has been the most difficult challenge you have encountered with your robot, and how did you use the design process to address the problem and implement a solution?

Corbin started. "Our greatest problem was that our robot sometimes froze at competitions, and we didn't know what was causing it."

Please don't try to explain our mess with The Best, thought Kenzie.

"We spent days discussing potential causes, studying the mechanisms on the robot, and trying to replicate the problem on our practice field," continued Corbin, "but we weren't making any progress."

"I tried changing the gear ratios on the motor that raises the tray to stack cubes," said Emma, "in case the robot was stopping because it simply didn't have enough power to keep going. But this didn't work. It turned out that, in this case, the amount of torque didn't matter."

"Even so, we thought it was important to include drawings and detailed descriptions in the journal of the changes that Emma made so as to document what we had tried," said Kenzie.

The judge in the center nodded his head.

"It turned out," said Corbin, jumping back in, "that the problem ended up being with the programming." Although this wasn't entirely accurate, it was much better than trying to explain that a crazy team was attacking their robot by hacking the field system. "We added some new code earlier today that, so far, appears to have fixed the problem. But we're still in the process of evaluating."

The judges asked a couple of other questions about their strategy and how they felt they were doing at today's competition. Then, after making a few notes, Tiffany announced, "Thank you all very much. Best of luck to your team in the remainder of the tournament today."

With that, the Robonauts left the interview room.

"How do you think we did?" asked Emma.

"Not good," Corbin said. "On that first question, we basically told the judges that we had a problem, that we didn't know what to do about it, and we took a last minute shot in the dark this morning to try to fix it."

"I don't think it's *that* bad," said Kenzie. "We had a problem and had to try *something*. And the documentation they have in our journal is helpful because it shows the details of what we tried and the process we went through."

"I do wish we had mentioned how we're using the vision sensor."

"Well, for better or for worse, what's done is done. And next time we'll make sure to point out the sensors," said Emma.

* * *

Now that the qualification matches had ended, representatives, one from each team, had lined up with their robots by ranking for alliance selection. The Robonauts, who had only lost two of their qualification matches, ended up ranking third, and Kenzie was proud to represent her team. The Best lost more of their qualification matches than usual due to the damaged intake, but thanks to Nora and Kyle's willingness to cheat, they ended up ranking #5. Starting with the highest ranked team, the ref asked, "Who would you like to choose for your alliance partner?"

As is often the case in alliance selection, the #1 team chose the #2 team, forming an alliance that would be very difficult to beat in the elimination matches.

It was now Kenzie's turn to select an alliance partner. While there were some excellent teams immediately below them in ranking, none of them had strengths that particularly complimented the Robonauts. More often than not, Kenzie had in mind ahead of time which team she would pick. In fact, she usually had a second pick selected in case her first pick declined or was already chosen. But today she was unsure of which team she should select.

"May I walk down the line?" asked Kenzie. A head nod from the ref indicated she could proceed.

As she passed each robot, Kenzie recalled how each had performed in the qualification matches. Nora, who was representing The Best, stood there expressionless as Kenzie passed by.

Is that a tear running down Nora's cheek? Kenzie wondered to herself, stunned. Regardless, there was no way that she was going to pick The Best.

Kenzie continued walking until she came to team #12, where she found the team representative for the Sonic Boomers. *No glasses*, thought Kenzie, *so this must be Loren*. She hadn't seen the Sonic Boomers much that day, but she remembered how well they had worked together at Tumultuous Towers. Goliath might not be the largest robot around, but the Sonic Boomers could play defense well. Maybe they would be a good pick? Glancing toward the bleachers at Emma and Corbin for confirmation, Corbin gave a head nod and Emma gave a thumbs up.

Walking back to her spot, Kenzie spoke into the mic and said, "We would like to choose the Sonic Boomers."

After locating the Sonic Boomers, the ref asked, "Do you accept?"

"Yes!" said Loren, "We graciously accept." He picked up his robot and moved to stand behind Kenzie, thrilled to be partnering with a team that had ranked so high and that he knew well from previously competing together.

The team ranked #4 ended up selecting a sister team way down the line.

When the ref reached Nora, she chose team #6. But team #6 promptly replied, "We respectfully decline." Moving on to the next team, Nora selected team #7. But they also declined. As did team #8. And team #9. And every unchosen team remaining in the line until Nora reached the team ranked #15.

Apparently The Best had made quite an impression on the rest of the teams at Skyscraper.

* * *

The elimination matches that followed that day were not as dramatic as the qualification matches had been. Although the Robonauts and Sonic Boomers played well together, this alliance ended up being eliminated early on. The Best and their alliance made it as far as the quarterfinals. But now that the elimination matches were over, it was time to present awards.

Since the Hobbits had been unable to compete for most of the day and another team had topped their Skills score, there was no hope for them to win any of the major awards. Nevertheless, Alex and Cory were excited to see how the Robonauts and other teams that they knew had fared. Sitting right next to them were the Robonauts, who were not feeling very optimistic about their chances of winning an award. They had not gone far in the elimination matches, they had only ranked #6 in Skills, and they were not very confident in the answer they gave to the judges' specific question earlier in their interview. On the positive side, they had received a second interview and had ranked #3 in the qualification matches.

Everyone watched as the MC presented the Skills, Design, and Tournament Champion awards. The Sonic Boomers ended up receiving the Judges award. So far, The Best hadn't received any awards.

But then it came time for the Excellence award. As the MC talked all about the Excellence award and the specific factors that go into determining the winner, Emma whispered to Corbin and Kenzie, "Y'all, I'm tired." It had been a terribly long day. "As soon as this ceremony is over, let's get packed up ASAP so we can get to the hotel."

Corbin and Kenzie nodded their heads in agreement and starting preparing themselves to get back to the pit to pack up.

The MC finally made it to the point of announcing the winner. "And the winner of the Excellence award at Skyscraper is... the Robonauts from Valley Creek High!"

Kenzie and Emma and Corbin all looked up in surprise. Not only were they awarded Excellence, this also qualified them for state! As the crowd clapped and cheered, the Robonauts stood up and made their way to the center of the gym.

As they received the trophy and had their picture taken, the crowd's cheering didn't let up in the least. And no one in the crowd cheered louder for the Robonauts than Alex and Cory and Euggie.

Chapter 28

Rap! Rap! Rap! came a knock on the door.

The receptionist behind the front desk looked at the entrance to the hotel to see who it was. All guests should have key cards for coming and going. Upon getting closer, he recognized the visitor. In fact, this visitor often came to the hotel at night to make deliveries.

"They're in Meeting Room A," said the receptionist, opening the door and letting him in.

Walking to the door of Meeting Room A which was propped open, the delivery boy called out, "Pizzas are here."

At once Corbin jumped out of his chair to get the pizzas. After paying for the delivery, he set the pizza boxes on a coffee table in the center of the room.

"OK, Luis," said Corbin to the Sonic Boomer holding the controller, "would you do the honors?"

From the across the room, Luis turned on his controller and drove Goliath over to the pizzas. The claw grabbed the top of the pizza box and raised it open.

"Pizza is served," said Luis, as everyone starting to dig in.

The meeting room in the hotel was made for comfort with a soothing fountain at one end of the room, a cozy fireplace at the other end, and soft, overstuffed chairs and couches in the middle. A double sliding glass door on the back side of the room led to a huge outdoor pool area surrounded by a tall, wooden fence. This room was the perfect place to gather and relax after a long, tiring day at Skyscraper.

Present at this meeting of The Resistance were Cory and Alex, the Robonauts, Euggie, and Luis and Loren from the Sonic Boomers. Sitting at the foot of Cory's chair and next to the fireplace was Alex. Right beside him, not by accident, was Kenzie. Everyone else was sitting on the couch or in chairs.

"Thank you all for coming," said Cory, as she brought this rogue meeting to order. "For starters, Nora and Kyle are staying in this hotel - and they don't know about this meeting."

"And I would like to keep it that way," added Euggie.

"So, I've asked Loren to keep watch outside for us while we talk."

With that, Loren got up from his seat and walked to the door. "Got ya covered," he said as he left the room, shutting the door behind him. Upon exiting the room, Loren sat down on a couch in the hotel lobby outside the meeting room.

The members of The Resistance spent the next half hour catching up on everything that had happened. The Robonauts learned how Nora and Kyle had been threatening Alex. Euggie and the Hobbits learned all that had occurred in the CRA robotics lab just last night and the great lengths the Robonauts had gone to in order to defeat The Best at Skyscraper. And they learned how Corbin had earned the new nickname, Bilbo. Luis sat there listening in disbelief as he heard for the first time how The Best had been freezing and frying robots.

"So, Luis," asked Cory, "I'm curious... what made you and Loren want to be part of this group?"

Luis sat up straight and pushed up his glasses. "Before this meeting, it was the way Nora treats people and how she goes around saying, *"We're The Best."* But now that we know everything that team has been up to, it sounds as if it's in everybody's interest to keep Nora and Kyle in line." Luis paused for a moment. "If they had not stolen the final elimination match at Tumultuous Towers, we would have won Tournament Champion and qualified for state along with the Robonauts." After finishing his answer, Luis sat back in his chair.

"Well, I for one am tired of being sidelined by Nora and Kyle," said Euggie. "Not only do their stunts cause trouble for everyone, but they keep me from being able to give 100% and have fun. Before today I had actually decided to quit robotics. But after seeing today how much fun robotics can be, I've realized there is absolutely no way I can quit. This is something worth fighting for."

"You do know," said Cory, "that you will be a key part of our efforts. You're our inside man."

"I'm up for it," said Euggie.

"For me," chimed in Alex, "I'm just tired of having to watch my back, constantly wondering what trouble they are going to cause next."

A few heads in the room nodded in agreement.

"How about the Robonauts?" asked Cory.

Without hesitation, Corbin was the first to speak up. "I'm mad at Nora and Kyle because of the time and efforts we've had to spend dealing with them instead of working to improve our robot. And what if we hadn't been able to update our code at the last minute earlier today?"

"I'm with him," said Emma.

As everyone looked at Kenzie to speak next, there was a brief pause. "After today, I have a different take on this. I'm just as upset at Kyle and Nora as the rest of you - and I want us to do whatever we can to prevent them from cheating. But today during alliance selection, I think Nora was about to break down and cry."

"Yeah, because we showed her they aren't *the best*," said Corbin. "And she couldn't take it."

"But I think there may be more to it than that," said Kenzie.

"Like what?" asked Alex.

"Well, I don't know about Kyle. But, I wonder if Nora is a bit broken. She isn't used to anyone hitting back at her. And she probably didn't know what to think after we hit back so effectively. For once in her life she wasn't able to just take control."

"I think it had to have been hard on Nora," said Luis, "when so many teams chose to reject her alliance invitation."

"Maybe she just needs someone to be nice to her and to help her realize that she doesn't have to act the way she does," suggested Kenzie.

"That's nice and all," said Corbin, "but I don't trust her. She's a bully and I don't think she's ever going to change."

"In any case," said Cory, trying to move things forward, "we'll need you, Euggie, to keep us informed of any ways that Nora and Kyle come up with to cheat. The moment you hear anything, please let me know and I'll pull everyone together to work on a counter-plan. And Corbin, we'll need you to keep deciphering and countering Kyle's code."

"Will do," said Corbin. "Here, Euggie, you'll need this." Corbin tossed him a small gadget. "This will let you connect a flash drive to your phone so you can send me any new code that Kyle writes as soon as possible."

Euggie caught the adapter and put it in his pocket.

"The rest of you," continued Cory, "watch for any messages I send out and be ready to jump in and help. Any questions?"

"Just one," said Corbin. "Does anyone else claim that last piece of pizza?"

With no takers, Corbin grabbed the last piece.

"Time for me to get Loren and head up to our room," said Luis, now that this meeting was drawing to a close. "I'll fill him in on everything." Picking up Goliath, he went out to get his brother.

As Luis walked out of the meeting room, he said, "Loren, you won't believe everything that's been going on." But Loren didn't hear him. He was sound asleep with this head tilted back on the couch.

"Loren!" said Luis, "It's time to go."

"Is the meeting over?" asked Loren, slowly raising his head and opening his eyes.

"Yes. And I've got a lot to tell you."

With one arm holding Goliath, Luis used his other arm to pull Loren off the couch. "C'mon. Let's go."

Back in Meeting Room A.

"And I better get up to my room before Kyle comes looking for me," said Euggie.

As the room continued to thin out, Emma decided she needed to speed things up. With a slight wink at Cory she said, "Wanna go for a walk outside?"

"Sure."

Now for Corbin.

"Hey, Corbin, I think they just put out a fresh batch of warm chocolate chip cookies at the front desk."

"Really?" asked Corbin as he got up and left to go check this out.

Emma and Cory followed him out, leaving Kenzie and Alex sitting there by themselves next to the fire.

As soon as Emma and Cory left, a text came through on Kenzie's phone. *Gordon will be jealous.* Followed by a string of laughing emojis.

Kenzie flipped her phone over so Alex wouldn't see Emma's message.

"It looks nice outside," said Alex. Wanna go sit out by the pool?"

He didn't have to ask twice.

Alex and Kenzie exited the room through the double sliding doors that led out to the pool area. Taking off their shoes and socks, they sat down on the edge of the pool, wading their feet in the water. They talked about everything that had happened at the tournament that day, about how much they both loved robotics, and how wonderful this trip had been.

"So, did you really see Nora crying today?" asked Alex.

"I couldn't be sure, but I think so."

"Maybe she was upset about losing."

Kenzie smiled, "Maybe. It was fun beating her, though, wasn't it?"

"Oh yeah," said Alex, "she had it coming."

"Do you think people like Nora can learn their lesson and change?" asked Kenzie.

Alex didn't answer right away.

"I guess it's possible," said Alex, "but Nora is one of the most stubborn people ever."

"Maybe she just needs someone to give her a chance and be her friend?"

"Maybe. But I agree with what Corbin said earlier. I don't trust her. And I don't want you to get hurt. So please be careful."

"Hey, Alex," came a voice entering the pool area. It was Corbin. And Emma and Cory were right behind him. "I'm heading over to the store across the street to look for some chocolate chip cookies. Wanna go with me?"

Either the cookies at the front desk hadn't been that good or, more likely, there were never any there to begin with.

"It's OK to go," whispered Kenzie. "It's getting late and I plan to head up to the room before too much longer. Thank you for tonight. I've enjoyed it."

"Me too," whispered Alex back.

"Yeah, I'll go," said Alex. He got up, slipped on his socks and shoes, and took off with Corbin on their cookie quest.

"And we're headed back up to our rooms," called Emma. "Wanna come with us?"

"Oh, I think I'll stay out here for just a couple more minutes," replied Kenzie. "It's nice out tonight."

"OK, try to be quiet when you return to the room cause I plan to crash as soon as I get back," said Emma as she walked away with Cory.

As Kenzie sat there by herself, lost in thought and swishing her feet around in the water, she suddenly heard her name. An unexpected visitor had come to see her.

Chapter 29

Earlier that evening at the hotel

Nora closed the hotel room door behind her and walked over to the bed. Sitting down and falling back, she revisited the events of the day in her mind. Nothing had gone as planned. How could her own robot have frozen? That was humiliating. And Alex's words, "Not the best", had cut deep.

Nora sat up, grabbed the remote, turned on the TV, and started flipping through the channels. Nothing good to watch. And nothing to get her mind off of everything that had happened.

Turning off the TV, she picked up her phone and texted Kyle. *Any luck figuring out what happened?*

No, came the immediate reply. *But I think the Robonauts must have messed with my code. Why else would their robot be the only one that was immune to our attacks today?*

Have you seen Euggie anywhere? texted Kyle again. *He hasn't been back to the room since returning from the tournament.*

Without responding, Nora threw her phone down beside her on the bed. Maybe a walk around the hotel would help clear her mind.

Nora got up and went out into the hallway, letting the door slam behind her. She took the stairs and made her way down to the lobby on the first floor. There she saw a young boy sitting on a couch in the lobby. And he was sound asleep. As Nora moved closer, she saw the fighter jet logo on his shirt and recognized him as one of the Sonic Boomers. She almost walked past, but stopped when she heard voices coming out of Meeting Room A. Moving just outside the room, Nora peeked through the crack between the double doors - and she couldn't believe what she found going on inside.

Although she didn't have a full view of the room, she could tell that the Robonauts and the Hobbits were both there talking about The Best. She listened as each person in the room listed off their grievances with her and Kyle. Except for Kenzie, who thought that maybe Nora just needed a friend. How sweet. But the greatest shock was when she heard Euggie's voice. Nora cracked the door a tad more so she could see for sure. And there she saw Euggie talking about how much of a pain the rest of his team was. And how he would continue to help this group in their efforts against The Best. Nora's blood began to boil.

Nora glanced back at Loren who was still asleep. If his task was to keep watch outside this meeting, he was failing his job miserably. As soon as it appeared this little get-together was about to end, Nora moved away from the meeting room and slipped out the main entranceway to the hotel. As Nora walked around outside, she continued to think about what her next move should be. Apparently Frodo getting fried hadn't been enough to deter Alex and Cory. And Euggie had turned out to be a complete traitor. As for those pesky Robonauts, it was high time she put them in their place - along with anyone else working with them. One thing was for sure. *Every last one of them would pay*.

After a while, Nora came to the back side of the fence that surrounded the pool. There she heard voices coming from the other side, one of which sounded like Alex. Moving up to the fence and peeking through a crack between the boards, she saw Alex and Kenzie sitting on the edge of the pool. She could hear them talking. Loud and clear. "Ugh!" thought Nora, listening to the two lovebirds, "that's the *last* thing I want to hear."

But then more people came out into the pool area. It was Cory and Emma and Corbin. Nora watched as Alex got up and left with

Corbin. Then she watched as Emma and Cory left, leaving Kenzie sitting there by herself. At that moment, a thought crossed Nora's mind and an evil smile formed on her face.

Pulling back from the fence, Nora made her way back into the hotel and to the pool.

* * *

Kenzie was still wading her feet in the pool when she heard the voice of someone coming towards her.

"Hi, Kenzie."

Looking up, Kenzie was startled to see Nora standing there.

Taking off her socks and shoes, Nora sat down beside Kenzie and put her feet in the water. "Whatcha doin'?"

"Oh, just enjoying the evening," Kenzie said warily, wondering why Nora was there.

"Me too. I love getting out at night. Your team did well today at Skyscraper."

"Thanks," said Kenzie, still wondering where this was going.

"It sure is a lot of pressure to get a robot built and working in time for competitions. Sometimes I think I would do *anything* to get rid of the pressure and know that our team would still win."

Kenzie knew exactly what she was talking about. Robotics *was* a lot of pressure and a lot of hard work. But Kenzie would never cheat to win.

"But I want to change things," continued Nora. "I mean, what good is winning if you haven't earned it?"

Kenzie couldn't believe her ears. Was Nora actually turning over a new leaf? Maybe there was hope for Nora after all. Maybe she should give her a chance?

Nora continued to explain how all of her team's efforts to attack opposing robots were Kyle's doing and how she had just been along for the ride. She was never comfortable doing the things she did, but felt like she didn't have a choice.

"What you all have done has been hurtful," said Kenzie, "but I guess I understand why you did it."

"Thanks Kenzie. It's good to have a friend like you who I can talk to."

And Nora did continue to talk to Kenzie. About school, about robotics, and about life. The more they talked, the more Kenzie began to feel as if she was starting to get to know Nora. It would be wonderful if they could reach a point where their new friendship would start to overtake all the ongoing problems.

"So what's up with you and Alex?" asked Nora.

"Oh, Alex is a great guy. And a great friend."

"Friend? I've seen you two together. Looks like it's more than that to me."

"Um, maybe," said Kenzie with a smile,

"Come on Kenzie, you can tell me," coaxed Nora. "I've seen you two together. You like him a lot, don't you?"

What harm could there be in talking with Nora about guys?

"Yeah, I do. He's sweet and we have lots in common."

At Nora's continued prodding, Kenzie went on to share more and more with Nora about her feelings for Alex. Then, as suddenly as the conversation had started, Nora announced that she was tired and it was time for her to get to bed.

"Ok," said Kenzie, "Good night."

As Nora got up and walked away, she pulled out her phone and stopped it from recording. Mission accomplished. This conversation could prove useful.

* * *

As Nora walked back to her room, she pulled out her phone to make a FaceTime call to Kyle.

"Hey," answered Kyle. He was lying on the bed in front of his laptop, obviously distracted and probably still trying to figure out why his code hadn't worked.

Nora got right to the point.

"I found Euggie," she said dryly.

Kyle just sat there working on his laptop.

"He was downstairs in the meeting room."

Still no reaction.

"He was meeting with the Robonauts and the Hobbits."

This got Kyle's attention.

"He was what?"

"You heard me. Euggie was meeting with them and they were sitting there discussing all kinds of things that you would find quite interesting."

As Nora relayed everything she had overheard from the meeting, Kyle became furious.

"That little traitor!" said Kyle. "I've been trying to figure out all evening how my code could have messed up and why it didn't have an effect on the Robonauts' robot." He couldn't believe Euggie was part of a blasted group working together against him. "When he gets back to the room, I'm gonna-"

"You're not gonna do anything tonight," cut in Nora.

"Oh, yes I am," said Kyle.

"No. You're not," Nora reiterated stubbornly. "They don't know that we're on to them. So, we're gonna teach them a lesson and make them sorry they ever messed with us. Especially the Robonauts. Here's what I wanna do..."

Kyle listened intently to Nora's scheme.

"Think you can do that?" asked Nora.

"Definitely," replied Kyle. "Getting things ready now."

* * *

Kyle was still lying on the bed working on his laptop when Euggie returned to the room.

"Any luck figuring things out?"

"No," said Kyle, "I'm not sure I'll ever know what happened today. But that's OK because I'm working on a new plan."

"Really?" said Euggie, "what's the new plan?"

"Oh, it's a surprise. And it will be *very* effective. I've been working on it all evening. Just to make sure I don't lose my work, I've got all the code backed up right here." Kyle held up a small blue flash drive, the same one Euggie had recently "borrowed" from him. Then setting it down on the night stand between the two beds in the room, Kyle continued to work.

Within the hour, Kyle had shut down his laptop, turned out the light, and gone to sleep. Or, at least it *appeared* he had gone to sleep. With one eye cracked open, he watched as Euggie quietly reached over to the nightstand and grabbed the flash drive. After a couple of minutes, Euggie returned the flash drive. Once Euggie had finally gone to sleep, Kyle opened his eyes and pulled out his phone.

<Kyle>
He took the bait.

<Nora>
Did you do what we discussed?

<Kyle>
Yes. When Corbin opens that file it will plant a virus on his laptop that will run in the background. It won't execute until the day of our next tournament. Once the virus is planted, it will send me periodic notifications so we know our plan is on track.

<Nora>
So how exactly does this virus work?

<Kyle>
It does two things. First, it will corrupt all the code files on Corbin's laptop, making them unusable. Then, if Corbin tries to download any changes to their robot's brain, it will render the robot useless.

<Nora>
When is the next tournament anyway?

<Kyle>
The first week after school starts back in the spring. It's a new tournament that was just added to the schedule this past week. And you'll never believe where it's being hosted.

<Nora>
Where?

<Kyle>
Valley Creek.

* * *

After sleeping late the next morning, the Robonauts loaded the van and began the drive back home to Valley Creek. Not far into the trip home, Coach Warren decided it was time to make an announcement.

"I want to let you all know that we have a special robotics event coming up in the spring the first Saturday after school starts back. We will be hosting a robotics tournament at Valley Creek. We're calling this tournament Soaring Heights. It's been an ongoing project for some time with our rep from REC Foundation, but robotevents.com posted this event a few days ago, making it official. I hope to have teams register from all over the area, both middle school and high school."

"That's great," said Kenzie, as if she were learning this news for the first time.

"Sounds like fun!" said Corbin.

"What a wonderful surprise!" said Emma, trying not to laugh.

"Most of the preparation for this event will be handled by an event team that I have put together. But, in addition to competing, I'm going to need your team's help. We have three brand new fields coming that will need to be assembled. One in a classroom for Skills and the other two in the gym."

"I think we can do that!" said Kenzie.

"You need to know that there is a basketball game scheduled in the gym on the Friday night before the tournament and we can't get started until after it's over. It will be a very late night."

Emma was already browsing robotevents.com to see the new tournament listing. "Looks like there are already several teams signed up that we know." "I see our team, Team Alpha, the Sonic Boomers, the Hobbits..." She paused. "And the teams from CRA."

"Speaking of CRA, I received an email this morning from Euggie," said Corbin, talking softer so only Kenzie and Emma could hear. He sent me some new code that Kyle wrote yesterday. It's in my email and I'll plan to check this out after we get back. Looks like our next battle with The Best will be on our home turf."

As the van continued its way back toward Valley Creek, Coach Warren had another issue he needed to discuss with the team. Bringing this up before everyone fell asleep would probably be a good idea.

"As much as you all focus on robotics, I'm guessing you haven't had much time to keep up with everything going on in the world."

"Like what?" asked Kenzie.

"Like this new Corona virus," answered Coach Warren. "According to reports, this new virus is making its way through a number of cities. Although it remains to be seen, there is quite a bit of talk about the need for people to start wearing face masks in public and maintain distance between each other. If all of this continues, there is no telling how it may affect robotics tournaments in the spring, or beyond. I just wanted you to be aware of this so that any major changes announced won't be a complete surprise."

While all of this sounded serious, the Robonauts didn't realize just how much of an impact this virus would end up having on robotics - and on their lives. They would soon find out.

Chapter 30

Two days had now passed since the weekend of Skyscraper and it was the last day of school for the fall at Valley Creek High. Since Valley Creek would be closed over Christmas break, Emma had talked Coach Warren into letting their team move the half-size robotics field from the school lab to the basement in her house. This way the Robonauts could continue working over the break and make sure they were prepared for Soaring Heights.

After their last math class of the year, Kenzie and Emma made their way down to the robotics lab to meet Corbin. As they entered the lab, they found Corbin already sitting on the field with tools in hand, disassembling the last section of the field perimeter.

"This is a great idea," said Corbin as the girls came in. "If we had to stop doing robotics over the break, I wouldn't know what to do with myself. I think my mind would probably explode."

"Not a bad idea, huh?" Emma said, proud that she had engineered this arrangement. "The programmer can thank the builder at any time since the programmer obviously needed the builder so he could continue his work over the break."

Corbin responded by pointing backwards over his shoulder with his thumb towards the back of his shirt. It read, *Be nice to the programmer. Santa is watching.*

"I still need a thank you," joked Emma.

"Thank. You." said Corbin, fully pronouncing each word. "Here ya go."

He handed her a screwdriver.

"All the small hardware goes in this plastic bin," continued Corbin as he pushed the container towards her. "And all the field elements and cubes go in those boxes by the wall. We can stack the field tiles beside them."

"So, have you had a chance yet to look at the code Euggie sent?" asked Kenzie, as she started pulling up the tiles one by one.

"Not yet, but hopefully I can do this tonight or tomorrow night," replied Corbin.

"Not tonight," said Emma, as she worked to pull the towers loose from the field.

"Why not?" inquired Corbin.

"Because we have our end-of-the-year robotics party tonight."

"Party?" asked Corbin and Kenzie at the same time.

"Yes, party." replied Emma. "At my house. Because there is no way I'm gonna be able to get this whole field set up again all by myself." Setting a field up took much longer than it did to break one down. Emma looked up with sad eyes that said, *please?*

"I'll make you all supper and pick up a six-pack of Dr. Pepper?"

"Whaddya think, Corbin?" said Kenzie.

"I think the builder can thank the programmer for all the time and effort that the programmer is about to give to the builder."

"I think that's a yes," said Kenzie, laughing.

"Great!" said Emma. "Come over at 7:00pm tonight and we'll get everything set up."

"Sounds like a plan," said Kenzie.

"I still need a thank you," said Corbin.

"Thank. You."

Shortly after the Robonauts finished disassembling the field, Corbin's parents pulled up to the school in their SUV to help transport it to Emma's house. With all three Robonauts working, it took several trips each to get everything loaded in the vehicle and to be unloaded again in Emma's basement.

“Thanks for moving the field over here,” said Emma, as Corbin got back in his parent’s SUV. “See you in a few hours for our robotics ‘party’!”

As the car pulled out of the driveway, Corbin yelled out the window, “Don’t forget the Dr. Pepper!”

Chapter 31

With the whole team working together, it took the Robonauts two hours to reassemble their half-size field. Emma's basement would now be their new lab for the next several weeks. After working up an appetite, the Robonauts sat on the floor to the side of the field so they could dig into their food and have a team meeting.

As promised, Emma had "fixed" a meal. If Pietro's to-go counted as fixing a meal. Since pizza was becoming way too common, Emma decided to try the new takeout special, a four foot Italian sub that the team could munch on all evening along with a huge basket of hot wings. And of course, the obligatory six-pack of Dr. Pepper.

"So," asked Emma, "was it worth it?"

"Absolutely," said Corbin, "I like the wings. You ought to try one."

"Aren't they really spicy?" asked Emma.

They *were* called hot wings for a reason.

"Not terribly spicy." Corbin glanced at Kenzie, trying not to smile. "Try one, you'll like it."

Emma picked up the basket of hot wings and sniffed them. But then she set them down. Maybe she would try one later.

"So, what improvements do you all think we need to make on Columbia before Soaring Heights?" asked Emma.

“I think we need to figure out how to stack a few more cubes,” said Kenzie. “Several robots at Skyscraper were stacking at least nine to ten cubes. If we want to be competitive at Soaring Heights, we need to be able to stack at least this many.”

Corbin nodded his head in agreement since his mouth was full of food.

“To do this,” said Emma, “I’ll need to make our tray longer.

“Just make sure we don’t fall into the same trap that Monstrosity did,” said Kenzie. “If the tray gets too heavy with more cubes, we don’t want to end up tipping over.”

Emma reached over and picked up a small dry erase board to start making a list.

“And,” said Corbin, swallowing his food, “we’ll have to make sure the longer tray can still fold down inside the initial 18” x 18” x 18” size limit.”

“And still pop out cleanly,” added Kenzie.

“The intakes will need enough power to pull the extra cubes in,” said Corbin.

“So,” said Emma, “here is what we have so far.” She read off the list she had put together. “Did we get all of our extra parts from the lab?”

Corbin gave a thumbs up since his mouth was again full of food and his Dr. Pepper was in his other hand.

“Corbin, have you looked at the code yet that Euggie sent?” asked Kenzie.

He shook his head no before reaching over and grabbing his laptop, which was sitting a few feet away. He opened the screen and set his drink down, motioning for the girls to come look.

As everyone gathered around, Corbin pointed to the email he had received from Euggie. It read, *Kyle says he is working on a new plan. Doesn’t sound like it can be anything good. Code is attached.*

“Looks like you have your work cut out for you,” said Kenzie. “How about you dig into this code while I help Emma put together a new tray extension?”

“Sounds like a plan,” said Corbin. He copied the file that Euggie sent to his desktop and put his mouse on top of it to open it.

Pausing, he looked up at Emma. "So, are you gonna try one of those wings?"

"Oh, why not," said Emma, grabbing a wing out of the basket and taking a bite just to satisfy Corbin. Kenzie and Corbin watched as she chewed slowly. Emma's eyes grew wide and her face started turning red. She whipped around, grabbed a napkin from behind her, and spit it out. "I thought you said they weren't that hot!"

Kenzie shrugged her shoulders. And Corbin sat there with a smile on his face - until Emma hurriedly got up to grab her drink from across the room. As Emma got up, her foot hit Corbin's Dr. Pepper, spilling it all over the floor right beside Corbin's laptop and splattering a good amount on the keyboard.

Corbin picked up his laptop and flipped it upside down to keep as little soda as possible from soaking in. Then he grabbed a handful of napkins and wiped off the keyboard the best he could. Meanwhile, Emma was across the room guzzling several swallows of Dr. Pepper.

"Umm... that can't be good," and Kenzie, not knowing what to do.

"Ya think?" said Corbin, still holding his laptop upside down.

"I'm so sorry!" said Emma, as she set down her can of soda.

"It's OK," said Corbin, setting a pile of drenched napkins off to the side. "Want another wing?"

Right after Emma gave Corbin a look saying, "You're crazy!", the carbonation from guzzling all that Dr. Pepper rose up and Emma let out a deep "Burrrrrrrrrrp!"

Corbin and Kenzie rolled over on the floor laughing.

Emma, who was actually proud of herself, grabbed a few more napkins and started to wipe the rest of the soda off of the floor.

"Is your laptop gonna be OK?" asked Kenzie, pulling herself together.

"I think so," replied Corbin, sitting up and flipping his laptop upright again. "But to be safe, I'm gonna stop and backup all of my files while it's still working." Corbin pulled a flash drive out of his backpack and inserted it into his laptop. As soon as his files were backed up, Corbin removed the flash drive and set it out of the way in one of the towers on the field.

The keys on the laptop were still working, but many of them were starting to move slower than usual from the sugar in the soda that was starting to harden. By tomorrow morning, the keys would probably be frozen in place.

I should be able to get the keyboard replaced in the next couple of days," said Corbin. "There is a little computer repair shop near my house and I'll stop by there first thing in the morning to see if they can help."

"Can you still take a look at Kyle's code tonight?" asked Kenzie.

"Sure can," said Corbin, "things will just go a bit slower since the keys on the laptop aren't moving as fast."

Locating the file that Euggie had sent him, he double-clicked to open it. When he did this, the laptop appeared to freeze up for a few seconds.

"Try it again?" said Kenzie.

But trying again just did the same thing.

"Oh well," said Corbin, "we can work with this again once I get the keyboard fixed."

The Robonauts decided to turn on some music and spend the rest of the evening working on Columbia's new tray extension before calling it a night.

Chapter 32

"Phase 8 completed," said an electronic female voice as each word zapped onto the center of the huge 36" screen. Kyle had been gaming in his room for the last 2 hours. Setting the controller down, he leaned back in his chair and stretched.

Kyle's room looked like the lair of a hacker. Shades drawn. Dimly lit. Posters all over the walls with slogans boasting that hackers ruled the world. And, of course, the monster computer system with a glowing keyboard and speakers with the thumping bass.

Kyle's phone which was laying on the desk suddenly lit up. It was Nora. Her text simply asked, *Any updates?* She had been asking this every day since Skyscraper. As Kyle sat there trying to come up with another way to say *not yet*, an email alert came up on his screen. This was what he had been waiting for. This email was from the rogue process which was now running incognito on Corbin's laptop. The alert read: *Virus will execute in 31 days, 2 hours, 37 minutes.* He finally had a substantial update for Nora.

<Kyle>
Yes, just received confirmation. Corbin has run the file and his laptop is now infected. All his code files are scheduled to be corrupted at 7:00am, the day of Soaring Heights.

Chapter 33

The computer repair shop was able to replace the keyboard in Corbin's laptop the same day he brought it in, allowing the Robonauts to continue working without any further delays. Indeed, the team worked hard over the following weeks to prepare for Soaring Heights. They only took a few days off to celebrate the Christmas holidays and the New Year. By the time school started back, Columbia was collecting and delivering up to ten cubes with reasonable consistency, although a perfect stack was never a given. Sometimes a cube would sling off the top of the tray when the robot was turning. Occasionally a cube would hang up when entering the new tray extension, preventing additional cubes from being collected. And sometimes the stack of ten cubes wouldn't land flat in the scoring zone for whatever reason, but would topple over. With lots of practice, however, Emma was getting better at minimizing these types of problems.

Strangely enough, Kyle had stopped talking about his new idea and Euggie was never able to lay his hands on further code updates to smuggle to Corbin.

It was now Wednesday evening, three days before Soaring Heights would commence at Valley Creek High. Cory had called another meeting of The Resistance and the Robonauts were all sitting on their field in Emma's basement waiting for it to begin. Right at 8:00pm, a bell sounded and the video call came through on Corbin's laptop. As Corbin accepted the call, the light on his webcam turned green and an image of Cory and Alex came up on the screen.

"Hey y'all. What is that?" said Cory, seeing Columbia sitting behind them with its new humongous tray.

"Oh, that's nothing," laughed Emma as she got up to move the robot out of viewing range. Even though The Robonauts were working with The Resistance, they would still have to go up against the Hobbits at Soaring Heights. No need to give away any unneeded information ahead of time.

"And how have you all been?" asked Kenzie, knowing full well how Alex had been since they had texted most every day since Skyscraper.

"Doing good," said Cory, "just a bit uneasy about any surprises Nora and Kyle may have up their sleeves."

Another bell sounded. It was Euggie. "Hey everyone."

Then another bell. It was the Sonic Boomers. "Hey... on the road... may get cut off..." Luis' tinny voice came in and out, with a flickering image of the two brothers leaning over their phone. Then just as predicted, their call dropped.

"I'll catch up with them later tonight," said Cory. "So Euggie, any new info from Nora and Kyle?"

"No, not really. Kyle keeps saying that his plan is ingenious and no one will able to stop our team from winning. But there haven't been any new code files that I know of."

"Even stranger," said Corbin, "is that the code file you sent wouldn't even open."

"That doesn't make any sense," said Cory.

"Maybe it was a bad copy coming from the flash drive?" speculated Corbin. "In any case, all we have to go on at this point is the original code. I think our best defense is the counter-code I wrote that makes a target robot invisible to Destroyer. We'll pass this code on to all the teams in The Resistance. At least it won't be possible for any of the robots in this group to be frozen or fried."

"That would be wonderful!" said Cory.

"Yeah," agreed Alex, "we don't ever wanna go through that again."

"So, how is Frodo doing?" asked Emma.

“Much better,” said Alex. “Several ports on the brain were damaged and unusable. At first we thought we were going to have to replace the entire brain. But then we realized we could fix things by moving all of our motors and sensors to undamaged ports. We were fortunate to have enough spare ports to be able to do this. Moving all of our cables around and updating our programming ended up taking the better part of four days. If the entire brain had to be replaced, this would have taken us much longer.”

“We’re glad Frodo is doing better,” said Emma. “So, where do we go from here?”

“Cory, I’ll send you the counter-code that will make Frodo invisible to Destroyer,” said Corbin.

“And, I’ll forward it on to the Sonic Boomers,” said Cory.

“I’ll keep watching for any new info from Nora or Kyle,” said Euggie.

“Our plan,” said Kenzie, “is to be at Valley Creek on Friday night because we have to help get everything set up for Saturday. We’ll be in touch if anything new comes up.”

With no more business to discuss, the call ended and the Robonauts decided to call it an early evening. Even robotics students sometimes needed to take a break and get a few extra hours of sleep. They would need it going into this weekend.

Chapter 34

The day before Soaring Heights

Kenzie's alarm went off at 6:30am Friday morning, her usual time to get up and get ready for school. She couldn't believe that Soaring Heights was only one day away. As she rolled out of bed and started getting ready, she picked up her phone from her night stand. There were thirteen new messages from Emma and Corbin. Looking closer at the texts, she saw what was up.

1:02am <Corbin>
K and E, are you all up?

1:10am <Corbin>
VEX rolled out a major firmware update yesterday. We need to update Columbia and test our programming ASAP.

1:37am <Corbin>
Just finished reading the release notes. This could be big!

1:49am <Corbin>
Text me as soon as you can.

5:01am <Emma>
What's going on?

5:38am <Corbin>
There you are, finally!

5:38am <Emma>
So sorry, some of us DO have to sleep ya know.

5:38am <Corbin>
We've gotta update Columbia.

5:38am <Emma>
OK.

5:38am <Corbin>
Soon.

5:39am <Emma>
How soon?

5:42am <Corbin>
Can I stop by before school? 7am?

5:53am <Emma>
I guess so. But I'll be upstairs getting ready, so come in the basement door. Kenzie, can you stop by too, in case you need to update the journal with whatever Corbin's doing?

6:32am <Kenzie>
Yes, but I'll have to move fast to make it by 7am. This better be good!

* * *

It was just after 7:00am when Kenzie arrived at Emma's house. Finding the basement door cracked open, she walked in to find

Corbin saying "Yeeees!" and shaking his fist in the air as if he had just won the final elimination match at a tournament.

"You're not gonna believe this," said Corbin as he started dancing around in celebration.

"Are you sure you're not just sleep deprived?" asked Kenzie. "Or maybe you've had a few too many Dr. Peppers this morning?"

"I'll put it like this," said Corbin, holding still for a moment. "Our problem with The Best is solved!"

And he started dancing again.

"Whoa boy, settle down," said Kenzie, "You're gonna have to explain this."

Corbin grabbed the controller and started driving Columbia around the field.

"This is Kyle's program running. When I press the button to attack a robot, look what happens."

He pressed the button.

Kenzie focused on the robot to see what it would do. "Um, nothing happened."

"My point exactly!" said Corbin. "The firmware update patched the problem."

It was now starting to make sense to Kenzie. "You mean, Nora and Kyle can't freeze robots anymore?"

"Not after Destroyer's brain gets the latest firmware upgrade. And not only that, but Nora won't be able to fry robots either. Kyle's ability to target and attack other robots is completely G.O.N.E."

This was worth calling Alex.

"Hello?" came Alex's voice from Kenzie's phone.

"Alex, I thought you would wanna hear this."

"Problem solved!" yelled Corbin from across the room.

"Huh?" said Alex.

"The problem with Nora and Kyle freezing and frying robots is now history," said Kenzie.

"That's great!" said Alex, "but how?"

Walking over to Kenzie and her phone, Corbin explained. "VEX put out a firmware update last night that included a ton of security enhancements. Once a robot gets this update, it will no longer be possible for Kyle's code to find and target other robots over the field system. So, he can still run his malicious code, but it won't be able to do anything."

"How do we know Destroyer will be updated with the latest firmware?" asked Alex.

"Inspection," Corbin promptly replied. "All competing robots have to pass inspection. And one of the requirements is having the latest firmware."

"That's exciting!" said Alex.

"I bet you're not quite as excited as Corbin over here," said Kenzie, as she watched Corbin break into another dance.

"I'm headed on to school," said Kenzie to her dancing friend. "I'll add to the journal that you did more testing this morning and confirmed that our problem is fixed."

Kenzie walked out of the basement and headed towards Valley Creek, still talking to Alex.

"So, are you and Cory ready for Soaring Heights?" asked Kenzie.

"Almost. I just brought Frodo into the lab and I'm here getting it packed up in our crate along with everything we need for the tournament. And then I'm locking it shut so you-know-who won't be able to mess with it."

"Oh yeah, I bet keeping Frodo away from Kyle and Nora is a pain."

"It wasn't so bad over break because I took Frodo home with me. But ever since school started back, I've had to keep it locked up when we're not around."

"You know," said Kenzie, "on that night at the hotel, after Skyscraper, Nora came to see me and she was actually being friendly. I tried to give her a chance. I was nice to her and am hoping that maybe she is finally softening up a bit."

"I hope you're right. But I wouldn't get your hopes up. We still need to keep a sharp lookout for any more stunts that she and Kyle try to pull."

"Yes, of course."

"Hey I gotta run. See ya early tomorrow morning."

"I can't wait," said Kenzie, with a smile.

Chapter 35

Walking down the hall of CRA, Kyle made his way to the robotics lab. He still had 45 minutes until his first class started. For once he was getting everything packed up early for the weekend tournament rather than waiting until the last minute. As Kyle approached the door to the lab, he saw Alex inside through the door's window. Alex was sitting on the opposite side of the room facing the other way. He was talking into his phone on the table beside him while starting to pack for Soaring Heights. Quietly, Kyle cracked the door and listened. And boy did he get an earful. Kyle had not yet discovered the new firmware update that had been rolled out. But he heard Alex and Kenzie talking all about it, loud and clear. This new firmware update could ruin everything!

As soon as Kyle headed into the lab, Alex ended his phone call, closed the crate, and locked it shut. Not wanting to stay in the lab with Kyle any longer than he had to, Alex got up and left the room.

Kyle went to the cabinet with the skull and crossbones and started searching through it. It only took a minute to find the small storage bin he was looking for. From inside the bin, he pulled out a second robot brain. This was a spare he kept in case it was ever needed. And it was definitely needed this weekend at Soaring Heights. Even though the new firmware update on Destroyer's current brain would be required at inspection, no one would know if he switched out to this second brain *after* inspection. This second brain would *not* have the latest firmware installed. Without the latest firmware, The Best would retain the ability to freeze or fry robots as they had been doing all season long.

Pulling out his phone, he started to text Nora. *You aren't gonna believe what's happened.* But then Kyle paused for a moment. Pressing and holding the backspace, he deleted the words he had just typed. His day would be much more pleasant if he didn't tell Nora about all of this until later.

As Kyle placed the second brain in a secure compartment within The Best's crate, his phone lit up. This time it was another automated email alert from the virus on Corbin's computer. The alert read, *Virus will execute in 23 hours, 22 minutes.* At least this part of Kyle's plan was still intact. Come 7:00am Saturday morning, all of Corbin's code files would be destroyed - right before the tournament.

Chapter 36

6:30am Saturday morning, the day of Soaring Heights

Kenzie was finally getting a few hours of much-needed sleep. But for some reason she couldn't get comfortable. She tried several times to roll over onto her side, but something kept getting in her way. As she groggily opened her eyes and sat up, she discovered the problem. She was hitting a tower. That was when Kenzie remembered where she was. She was lying on one of the new competition fields that she and Emma had finished setting up only a few hours ago in the Valley Creek gym. And this stupid tower beside her was preventing her from sleeping.

Last night's basketball game had delayed the Robonauts in setting up for the tournament for over two hours. So, they ended up working into the early morning hours to get everything ready in time. Although the field perimeters and floor tiles were setup by midnight, the field elements still needed to be put in place. Corbin had bailed out at 11:00pm. And Coach Warren left around midnight, but he had agreed to let the girls stay and keep working. By the time they had finished, it was after 1:00am. Rather than go home this late and return only a few hours later, the girls reasoned that they might as well just crash right there on the field.

Kenzie glanced over to the other side of the field to see Emma, curled up in a ball and sound asleep up against the field perimeter. She looked at her phone to see what time it was. 6:32am. Coach Warren would be there with his event team right at 7:00am, and teams could start arriving soon after. Inspection was scheduled to open at 8:00am.

* * *

6:35am

The CRA bus pulled out of the parking lot right on schedule. As always on tournament days, the bus was filled with noisy conversation and excitement about everything that would happen in the coming hours. Sitting in their usual back seats were Nora and Kyle. Noticeably absent from the back of the bus, however, was Euggie. He was sitting by himself several seats up from them, behind Cory and Alex. This day was already going to be awkward for Euggie and the Hobbits. But after the bus started rolling down the road, Nora decided to make things worse.

A wad of paper launched from the back of the bus landed right between Cory and Alex. Picking up the paper and un-wadding it revealed the word, LOSERS.

"Hey Hobbits, are you ready to lose today?" called Nora from the back of the bus.

"Oh boy, here was go," said Cory to Alex without turning around.

Both just sat there, ignoring her.

"Hopefully your robot will be able to stand up to the competition today better than it did at Skyscraper," Nora taunted, turning up the pressure and standing with her hands on her hips.

Not able to stand it any longer, Cory turned around and looked directly at Nora. She stood up and stuck out her tongue at her while cocking her head sideways and going cross-eyed. Then she sat back down.

Euggie smiled with the realization that something so simple would completely infuriate Nora. It took so little.

Nora shot back, "It would be a shame if your robot had *more* problems today."

"Oh, shut up Nora!" came a voice from some student mid-way up the bus.

"Yeah!" agreed another student across from where the Hobbits were sitting.

It was time to calm Nora down. "Better save your energy for the tournament," whispered Kyle. "We'll show them once and for all who is really the best."

Nora sat down in a huff with her arms crossed.

"I'm guessing Nora's mood isn't going to improve any today," whispered Cory to Alex.

"Yeah, no kidding. I've worried all week long about what she might try to do to us or Frodo. Ever since we came back to school, I've made sure to keep Frodo locked up in our crate. No one can unlock our crate without this." Alex pulled a small key out of his pocket and handed it to Cory.

Just then another wad of paper from behind bounced off of Alex's head before landing on the floor.

This was going to be a difficult day.

* * *

6:37am

Still sitting on the field in the Valley Creek gym and a bit more awake now, Kenzie texted Alex.

<Kenzie>
Are you all on your way?

<Alex>
Yes, we should be there just after 7:00am.

<Kenzie>
Everything is in place and ready to go. We're still in the gym. Never left last night. Any new info from Euggie?

<Alex>

Nothing new from Euggie. He's excited about the firmware upgrade though. Hoping this will put an end to Nora and Kyle's stunts. The only thing new here is that Nora is being more of a pain than usual this morning.

Without getting up, Kenzie pushed a long row of cubes across the field towards Emma, gently nudging her. "Hey girl, wake up!"

Emma's eyes popped open and she sat up quickly. "What time is it?"

"6:40am," said Kenzie. "Alex says that CRA will arrive here just after 7:00am. He also said that Nora is really being a pain today. I'm going to run back to my house to grab my competition shirt. Want me to grab your shirt on the way?"

"Yes, please," said Emma, tossing her the key to the basement door. "It's upstairs in my room hanging on the door knob."

"OK, text me if you think of anything else you need."

On her way out the door, Kenzie saw a small minivan that had pulled in and parked. Piling out the side door were Jake and the rest of his platoon, dressed in camo as was usual for their competitions. They sure were here early. "Good morning, Team Alpha," said Kenzie. Jake responded with a salute as she passed by.

As Kenzie continued walking, she noticed that the sky was turning gray and a cool breeze was starting to blow. Tree leaves were shimmering in the wind. A coarse rumble of thunder followed distant flickers of lightning in the sky. Picking up her pace a bit, Kenzie opened her weather app. The local radar confirmed what she was already seeing. A huge storm was brewing.

Chapter 37

6:48am

Emma held the door open as two members of Team Alpha and their coach filed into the Valley Creek gym carrying equipment destined for their pit.

"Welcome!" said Emma, "You all are the very first team to arrive. Our workers aren't here yet but you can go ahead and get set up. The pit area is spread out around the back of the gym. Just find the table labeled with your Team ID."

"Yes, ma'am!" said one of the little soldiers.

As Team Alpha was delivering equipment to their assigned table, their two-way radios went off.

"This is Alpha 1. Bringing in our last container of supplies."

"Copy that Alpha 1," responded Alpha 2.

This gave Emma an idea.

A few moments later Emma opened the gym door for Alpha 1.

"Jake," said Emma as he came through the door, "we need to talk."

Team Alpha was going to be quite helpful today.

* * *

7:00am

The gym was now coming alive with activity. Teams were arriving and people were scurrying around, preparing for the tournament. Coach Warren and the event team had arrived, including Curly who was donning his usual red baseball cap. While Coach Warren started inspecting each field that had been set up, Curly began powering on the computers at the control table and cranking up Tournament Manager. The rest of the event team began making their way around the gym to take care of the long list of tasks that had to be completed. They made sure that the inspection area was set up, that the registration table was manned, that lines for the queuing tables were clearly marked, and that someone was in place to host and assist the judges as they arrived. Outside, the storm was picking up and the first signs of rain were appearing.

"Good morning, Luis and Loren!" said Emma as she opened the door for them. "Pits are at the back of the gym."

The Sonic Boomers came through carrying Goliath.

"Thanks. It looks like The Best is arriving," said Luis, pointing backwards with this thumb towards the parking lot.

Looking out the gym door, Emma saw the CRA bus pulling in.

"We'll keep an eye on them," said Emma, as Luis and Loren continued on to find their pit.

Catching Jake's attention from across the gym, Emma gave him a firm head nod. After returning the same head nod, Jake rounded up the rest of his team and led the way as they marched over to Emma.

"Reporting for duty!" said Jake.

"Great! You know what to do," said Emma as she opened the gym door for Team Alpha to head outside on their first assignment.

* * *

7:03am

Kyle and Nora were starting to pull their heavy crate from the cargo area of the CRA bus when Alpha 1 and his platoon walked right up to Nora and stood at attention.

"May we assist, ma'am?" asked Alpha 1.

The expression on Nora's face said, *What planet are you from?* But as she saw the gathering storm, she relented. They did need help getting all their stuff inside as quickly as possible. It only took Team Alpha a minute to unload The Best's crate. Then, piling their tool chest and a few other containers on top of the crate, Team Alpha began rolling all of The Best's equipment up the sidewalk and towards the gym.

Emma held the door wide open for Team Alpha as they came through. Nora and Kyle followed, awkwardly passing by Emma without saying anything. After leading Nora and Kyle to their pit and delivering their equipment, Team Alpha then returned to their own pit. Except for Jake. Jake went to the bleachers and took a seat, not too far behind The Best. Looking at Emma who was still standing by the door, he gave her a thumbs up.

Emma returned the thumbs up and looked at her phone. 7:06am. And it looked like the storm outside was about to hit at any minute. Where was Kenzie?

Where are you? texted Emma. *Did ya get lost?*

* * *

7:07am

Kenzie was almost back to the school parking lot, picking up her pace as the rain was coming down harder. It was important that the surprise she was carrying for Emma not get wet. Ahead of her, she watched a small navy blue car as it pulled into the lot and parked. Then, with the driver's door opening, a huge black umbrella poked out. As Kenzie drew closer, she was surprised to see it was Gordon getting out of the car.

When Gordon saw Kenzie coming, he called out to her, "Need an umbrella?"

At that very moment, the bottom dropped out the sky and heavy rain started to pour. With no time to answer, Kenzie ran as quickly as she could to get herself and everything she was carrying under Gordon's umbrella. In the process she ran right into Gordon, almost knocking him over. But it only took Gordon a second to regain his footing.

"Thanks," said Kenzie, scrunching herself under the umbrella with Gordon and trying to stay dry. "That was good timing. Um... are you OK?"

"Yes, I'm OK."

Gordon was most definitely OK because there was no one else he would rather share an umbrella with. But, with his attention on Kenzie, Gordon didn't realize that his phone had fallen out of his pocket and to the ground. It was now sitting face down on the pavement, under the edge of his car.

A flash of lightning streaked across the sky followed by a loud crash of thunder.

"We'd better get inside," said Gordon.

With no disagreement from Kenzie, they began walking towards the gym, huddled together under the umbrella.

* * *

7:08am

"Let's get this inside," said Alex.

He and Cory had just covered their crate with a tarp when the hard rain began coming down. Euggie, who was avoiding Nora and Kyle as long as possible, was with them. Covered in ponchos, Alex and Euggie both had their hands free for pushing the crate forward. On the front side of the crate, Cory guided with the pull-rope in one hand and an open umbrella in the other hand. As they started moving, Alex happened to glance up. Through the rain and across the parking lot, he saw Kenzie. She was walking up the sidewalk and huddled under an umbrella. *Who was that guy she was with*?

* * *

7:09am

As Kenzie and Gordon moved under the awning outside the gym doors, Gordon closed his umbrella and shook the water off of it. Kenzie glanced at her phone, seeing Emma's message for the first time. She laughed and walked into the gym with Gordon. And there was Emma waiting for her.

"No, I didn't get lost," said Kenzie.

She handed Emma her shirt and then held up the surprise for Emma to see.

"You didn't!" squealed Emma with delight.

"I did."

Kenzie was the best friend on the face of the earth! She had apparently stopped by the little 24/7 gas station convenience store right across the road next to Pietro's, because she stood there holding a box of one dozen chocolate-covered creme-filled donuts. And had somehow managed to keep them dry!

"Sorry it took me so long. I thought we could use some breakfast," said Kenzie. "Then I got caught in the rain."

"And what are you doing here?" asked Emma, looking at Gordon.

"I thought I'd come out today to support you all."

It must have taken a huge effort to get out in this weather, thought Kenzie.

"Thanks for coming," said Emma. "We've gotta go get ready for the tournament. But you're welcome to go hang out in our pit. Just find the table that has our team name on it. Corbin should be here soon."

"OK, see you all in a bit," said Gordon, as he went off to find the Robonauts' pit.

As Kenzie and Emma walked out of the gym together to change clothes and find a place to eat their breakfast, Kenzie noticed Nora and Kyle busily working in their pit.

"Maybe we should have asked Gordon to keep an eye on Nora and Kyle for us," whispered Kenzie.

"Nah," said Emma, "I have it covered. Watch this."

As they stepped out of the gym and into the foyer, Emma pulled a two-way radio from her belt, an extra that Team Alpha had generously let her borrow. She spoke into it saying, "Alpha 1, Robo Em is taking a short break. Keep me posted if there is any change in the status of TB."

"Copy that, Robo Em," replied Jake's voice through the radio.

"TB?" asked Kenzie.

"The Best," replied Emma.

"And, Robo Em?" Kenzie burst out laughing.

"Hey, it works," said Emma. "It's the first call name that came to mind. And we now have Team Alpha working with us. They've agreed to keep me informed of the whereabouts of Nora and Kyle."

* * *

7:13am

Right after Kenzie and Emma left the gym, Alex and Cory came in, rolling their crate through the door. It took them a few minutes to find their pit and start unpacking.

"Do you have the key?" asked Cory, as she pulled the tarp off of the crate.

"No, I thought I gave it to you," replied Alex.

Cory checked her jacket pockets.

"Ugh, I don't have it."

"Don't look at me," said Euggie.

"I bet we left it sitting on the bus. I'll go back and get it," said Alex picking up his poncho again.

* * *

7:14am

"What time is Corbin's laptop going to blow up?" asked Nora.

"Shh," said Kyle, looking up from his laptop. "Voices carry in this gym. His laptop isn't *literally* going to blow up. The virus on his laptop will execute after 7:00am. Once the virus executes, it should only take a few moments for all of his code files to be destroyed."

"OK genius, what happens if Corbin doesn't turn on his laptop until after 7:00am? Because it's already after 7:00am. And it doesn't look like he is even here yet."

"Don't worry, after 7:00am the virus will execute immediately the next time he turns the laptop on."

Feeling like there were a pair of eyes on her, Nora glanced back to the bleachers behind her. But no one was there other than one of the little soldiers in camouflage that had helped them unload, just sitting on the front row. Whatever.

Meanwhile, Gordon was walking through the pit area, partially looking for the Robonauts' table and partially just taking in everything set up in the gym. Since Soaring Heights was his first competition, all of this was new to him. As he walked up and down the rows of tables, he found one table with no indication of what team was assigned. As he got closer, he realized the sign had simply fallen off and was lying under the table. Kneeling down to pick it up, he heard voices talking from a table a short distance away.

"Corbin and his team won't know what hit them after you fry Columbia," said Kyle.

Corbin? thought Gordon. He had no idea who these students were, but he decided he should stay down on the floor and listen.

"To make sure everything works right," continued Kyle, "I need to go find a place to do some work on Destroyer after inspection,"

"What's wrong with right here?" asked Nora.

"I can't do it here." He sighed, knowing that he would have to tell her sooner or later. "A VEX firmware update came out a couple of nights ago that messed up some of my programming, and this update is required at inspection. Once Destroyer is inspected, my plan is to swap the current brain out with this one."

He held up the spare brain for Nora to see.

"Swap brains? Right before a tournament?" asked Nora, rather loudly.

"Shhh!" said Kyle. "We don't want the world to know we're doing this!"

Kyle held up the spare brain that he planned on using. "This brain doesn't have the update and will continue to allow you to freeze or fry robots."

Suddenly Nora realized what Kyle was saying. Without swapping brains, she would not be able to attack Columbia or any other robots.

"Fine," snapped Nora. "But this had better work."

With that, Kyle grabbed his backpack and left to go find a good location for the brain swap.

After Kyle left, Gordon slowly peeked out above the table. Seeing the girl sitting there facing the other direction, Gordon got up and continued to walk on by.

Although Gordon didn't completely understand everything he had just heard, it didn't sound good. He needed to text Kenzie to let her know. It was at this point that he realized that he didn't have his phone. The last time he remembered having it was when he got out of the car. He must have lost it somewhere between the car and the school. Probably when Kenzie ran into him and almost knocked him over. By now his phone was most likely water-logged and useless.

Just then Corbin walked into the gym. As soon as Gordon saw him, he went over to alert him of what he had overheard.

Chapter 38

7:17am

Kyle made his way down a hallway of classrooms. Turning a corner, he found one that looked like it would work well - Room 108. Turning on the lights and looking around, he spotted an empty desk with a power outlet right beside it. This room would be the perfect spot. Kyle set his backpack down on the floor beside the desk and began unpacking it. First, he pulled out and unfolded an anti-static mat that he always used as a base when doing this kind of work. Then, on top of the mat, he laid out his laptop, the second brain, and all the tools he would need to make the swap. After one last check to make sure he had everything he needed, Kyle turned off the light and headed back to the pit area in the gym to meet Nora.

As teams continued to arrive, they were coming in soaking wet and bearing umbrellas, tarps, and whatever was needed to keep their robots dry. The storm made unloading a miserable experience. Picking up a mic on a stand in the center of the gym, Coach Warren announced, “We’re starting inspection earlier this morning than originally planned.” With the bad weather outside, teams would be slow in arriving. This would help keep things moving along and prevent the line from getting backed up at the last minute. “Inspection will now open at 7:30am and we encourage all teams to go through as quickly as possible.”

"I'm going to go get in line now to make sure I'm first," said Kyle as he grabbed Destroyer. Looking at Nora, he continued, "While I do that, you need to turn in our notebook and make sure we are registered."

"OK. Where is our notebook anyway?" asked Nora, looking around their pit table, but not finding it.

"Oh, I left it in my backpack. In Room 108. It's outside the gym, down the hall, and around the corner. It's easy to find," said Kyle as he headed off to claim first place in the inspection line.

Sure, leave it to me to go get the notebook, thought Nora. *Whatever.*

* * *

7:19am

After changing into their team shirts, Kenzie and Emma ended up going to their robotics lab. This was a good place for the girls to leave anything they didn't want to keep in the pit. It was also a good place to sit down and partake of the nutritious breakfast that Kenzie had picked up. As soon as they started eating, however, Emma's radio went off.

"Robo Em, we have movement," said Alpha 1. "Kyle has exited the gym. We followed him down the hallway and watched him go into Room 108. After a couple of minutes, he then exited the room and has just now returned to the gym."

"Copy that Alpha 1." replied Robo Em.

"Wow," said Kenzie, "those guys are efficient!"

"I know, right?" said Emma.

Kenzie paused. "Wait a minute. Room 108 is right next door to this room! I'll go see what he's up to." Kenzie set down her donut and slipped out of the robotics lab to check out the room next door.

A moment after Kenzie had left the robotics lab, Emma's radio went off again.

"Robo Em, Nora has now left the gym and appears to be headed to Room 108."

Kenzie was still in there!

Jumping up as quickly as she could, Emma ran out of the lab and opened the door to Room 108.

"Get back in here!" whispered Emma as loudly as she could.

"Why, what's up?" asked Kenzie.

There was no time for questions. Emma grabbed Kenzie's arm, pulled her back into the robotics lab, closed the door, and flipped off the lights. Emma put her finger over her lips, signaling for Kenzie to stay quiet. As the girls peered through the window in the door, they watched Nora walk down the hall and go into Room 108, right next to them.

"Robo Em, do you copy?"

Turning down the volume, Emma whispered into the radio, "Yes, I copy."

After a moment had passed, Kenzie and Emma watched Nora exit the room and head back down the hall, probably returning to the gym. Alpha 1 confirmed as much over the radio.

Relieved that Nora hadn't seen them and with their hearts still beating fast, they stood there trying to figure out what to do next.

"Eye of the Tiger" suddenly started playing on Kenzie's phone. It was a call from Corbin.

"Where are you?" answered Kenzie.

"In the gym. They just announced that inspection is opening early at 7:30am."

Kenzie pulled the phone away from her ear long enough to look at the time. 7:22am.

"Nora and Kyle are up to something in the room next to our lab."

"Yes, that's what I hear. Can you all get out here?"

"On our way," said Kenzie, ending the call.

"Hold up," said Emma, running back to grab the box of donuts. "We better take these with us if we are going to have any chance of actually eating them."

* * *

7:23am

When Kenzie and Emma returned to their pit they found both Corbin and Gordon sitting there. Cory and Euggie had also stopped by since they were still waiting on Alex to return with the key to their crate.

"Donuts!" said Corbin, noticing the box that Emma was carrying.

"Here ya go," said Emma as she set the box on the table and opened it .

As everyone at the table dug into the donuts, Corbin didn't waste a second getting down to business.

"Gordon overheard Nora and Kyle talking a few minutes ago and you wouldn't believe what they're planning."

"Something about swapping their brain after inspection," said Gordon, "so they could... fry robots? It didn't sound good."

"So that's what they are planning!" said Kenzie. "I just saw Room 108 where Kyle is set up to do the swap. What are we gonna do?"

"Well," said Emma, glancing over at Kyle who was first in line for inspection, "I vote we march down to Room 108, grab that second brain, and throw it out the window and into the parking lot."

"Or," said Corbin, as the gears started turning in his head, "maybe we can teach Kyle and Nora a lesson they'll never forget. We need to find out if anyone in The Resistance has a spare brain."

"I'm on it," said Cory.

Within seconds, Cory blasted a message out to everyone in The Resistance. While she was waiting for responses, she decided to check on Alex who had been gone now for 10 plus minutes.

You OK? Did you find the key?

Alex replied immediately. *The key wasn't in our seat, still looking for it on the bus.*

He needed to hurry. Inspection was going to start soon.

Just then, Cory got a reply from Luis.

"The Sonic Boomers have a spare brain we can use. They're bringing it to us now."

Almost immediately, Luis and Loren came running over to deliver their spare brain to Corbin.

* * *

7:25am

Across the gym from The Best's pit, Nora was watching all the activity going on at the Robonauts' table. They sure were working hard at *something*. But what Nora really wanted to see was Corbin's face after he turned on his laptop. This would be fun to watch.

As soon as Corbin had the Sonic Boomer's spare brain in hand, he pulled his laptop out of his backpack and turned it on.

"Emma, I need to make one quick change in our code on Columbia, then you can head over to inspection while I work on this spare brain."

Emma nodded her head at the plan.

But for some reason, Corbin's laptop was extremely slow starting up. And VEXcode was even slower loading. Fortunately, his code file did finally open. Corbin made the quick change to Columbia's code, recompiled, and downloaded the update to Columbia's brain. But when he did, something went terribly wrong. While the file updates appeared to transfer to the brain, the status bar turned red and the screen froze.

"What?" said Corbin under his breath. He quickly forced VEXcode to close and then tried to reopen his code. But when he did, he got an error message that no one ever wants to see, especially when opening code that is critical. *Error: Invalid File.*

"You all, we have a problem."

"What's going on?" asked Kenzie.

"Our code is gone."

"What?" asked Kenzie.

"It's just gone."

"Have you tried restarting?" asked Emma.

“Ha ha,” said Corbin dryly. “This is serious. Without our code and the required competition template, we’re not gonna pass inspection.”

As the Robonauts stood there in shock, not fully understanding what had just happened, the reality began to set in of what this could mean. Would they even be able to compete today?

Over in the inspection line, Kyle felt his phone vibrate in his pocket. Pulling it out, he found the email alert he had been waiting for. *Virus execution complete. Files destroyed.* The first part of his plan had been successful.

Chapter 39

7:26am

Nora sat at The Best's table across the gym, enjoying every moment of the drama unfolding in the Robonauts' pit. Glancing up for a moment, Kenzie noticed that Nora was sitting there watching.

"Do you all think Nora and Kyle are somehow responsible for this?" asked Kenzie.

"Probably," said Corbin, "but there's no telling. Right now I'm just trying to figure out what to do."

"Don't you have any backups of your files?" asked Kenzie.

For all of Corbin's talent in technology, keeping good backups was not his strong suit.

"No, I don't," said Corbin softly as he rested his head down on his hands, feeling helpless.

"Wait a minute!" said Emma. "Didn't you make a backup of all your files? You remember! That night that I spilled soda all over your laptop?"

Corbin's head shot up and looked at Emma.

"Yes!" said Corbin. He had not made significant changes to the code since that backup was taken. "The files are on my flash drive! Now where is that thing?"

Corbin started looking for the flash drive in his backpack.

"The programmer can be thankful that the builder spilled Dr. Pepper all over his laptop so he would backup his files," said Emma, as she watched with amusement.

There was no argument from Corbin on this.

"Now if we could just find it. It's not in my backpack."

"Unless it's been moved, my guess is that it's still in that tower on the field back in my basement," said Emma.

"We have to get it," said Kenzie.

Just then, there was a loud crash of thunder and the lights in the gym flickered off and back on a couple of times, a reminder of the storm and the pouring rain outside. Getting back to Emma's house would be a complete pain. Kenzie was already feeling pressure to secure a place in line to register their team and turn in their notebook. Emma still needed to finish preparing Columbia for inspection. And Corbin needed to stay right where he was since his plan to deal with Kyle's pending brain-swap was about to be set in motion. Yet, without the flash drive, none of these things would matter.

"I'll go get it," offered Gordon.

All three Robonauts looked at him. This would be *very* helpful.

"Do you know where my house is?" asked Emma.

"Approximately," said Gordon. "If you give me the address, I can find it."

"You do realize that it's pouring rain outside and you're gonna get completely soaked?" asked Kenzie.

"Yes," replied Gordon, "but I drove here and parked close to the gym. Between my car and my umbrella, I'll stay dry enough. Besides, a little water never hurt anyone. And I can be back in just a few minutes with it."

"Gordon you are a lifesaver," said Emma. "If you go get this flash drive for us, our whole team is gonna give you a huge kiss."

Corbin winced. Gordon blushed. And Kenzie gave Emma a look that said *Are you kidding me?*

"Or not," followed up Emma.

"It's OK, I'm glad to help," said Gordon. "I know this is really important."

After Emma had given her address to Gordon and Kenzie had given him the key to the basement door which she still had from earlier, Gordon took off as fast as he could.

As Gordon ran off to help the Robonauts, Kenzie found herself actually admiring Gordon. He had the firmest resolve of anyone she had ever met. He had come out to show his support today despite the storm. And now he was working with her team to help when it was very much needed. But there were more important things to think about right now. Kenzie grabbed the notebook and headed to the registration table.

Just then Cory got another text from Alex. *Found the key! Apparently it fell and was on the floor of the bus several rows down.*

* * *

7:29am

As Gordon ran out the gym door into the pouring rain, he was laser-focused on completing his mission. So focused that he wasn't paying attention to the steps he was taking, which one should always do, *especially* when running through heavy rain at top speed. Hitting an uneven segment of the sidewalk, Gordon tripped.

At the same time, a student in a poncho was coming up the sidewalk in the opposing direction, moving at a good rate of speed himself. This unlucky student ended up right in the path of Gordon's fall. An out-of-control Gordon fell smack dab into him, knocking them both to the ground. They fell backwards with their rear ends hitting the wet pavement. Gordon's umbrella had fallen off to the side and now he was completely soaked.

"Watch where you're going!" said the student.

"I'm so sorry. Hope you have a good day," said Gordon. And with that, he got up, picked up his umbrella, and ran off, refocused on his mission to obtain the flash drive.

As Gordon ran off, Alex looked backed at him. It was the same guy he had seen earlier with Kenzie.

When Gordon got to his car, he closed his umbrella and threw it in the back seat. Opening the driver's door, he started to get in when he noticed his cell phone lying on the ground. Quickly, he picked it up, got in the car, and shut the door. The display was dark around the edges where water had seeped in and the screen had a single crack in it running diagonally from top to bottom. But surprisingly, it came on and appeared to be working.

* * *

7:30am

Back in the gym, inspection was about to open. Kyle was first in line and chomping at the bit for things to get started. Kenzie was third in line at the registration table. Cory was back at the Hobbit's pit with Alex who had just returned with the key to their crate. Euggie was hanging out with the Hobbits. Corbin and Emma were still in their pit with the Sonic Boomers.

"I sure hope Gordon gets back with that flash drive soon," said Corbin. Indeed, much was riding on the success of Gordon's mission. Corbin needed enough time to follow through on his plan to deal with Kyle before Destroyer completed inspection. Then, there was the small matter of restoring Columbia to working order so the Robonauts would be able to compete. Unfortunately, it looked like time was running out. Corbin watched across the gym as the inspectors began sitting down at the tables and motioning for those in line to come forward and be inspected.

Right at that moment, a loud crash of thunder caused everyone in the gym to jump. Immediately after the thunder ended, the power went out - leaving everyone in the dark. Other than glowing phones, laptop screens, and red lights from several robot radios scattered around the gym, the room was completely black. After a brief silence, as everyone realized what had happened, activity slowly started up again. Surprisingly, activity in the gym continued to carry on with a fair amount of normality. Registration continued, although much slower as forms were filled out using the light from each person's phone. However, since it was incredibly difficult to inspect a robot in the dark, inspection was put on hold until the power was restored.

Could the inspectors not use their phone lights to inspect the robots? thought Kyle impatiently. But there was nothing he could do about it. So he just stood there waiting.

Now that it was much harder to see, or be seen, especially from across the gym, Nora decided to get up and move closer to the Robonauts. Walking around the perimeter of the gym, she sat in the bleachers, just close enough to keep watch on them without being too conspicuous.

Chapter 40

7:32am

Gordon pulled into the driveway at Emma's house. The rain hadn't let up at all. But this didn't matter. Since he was already soaked to the skin, he walked right through the downpour, made his way to the basement door, and went inside. The room was completely dark. Feeling along the wall, Gordon finally found a light switch and flipped it. But no lights came on.

The power must be out, thought Gordon.

But he wasn't going to give up.

With the small amount of light that his phone was still putting out, he could only see a few feet in front of him. Gordon took one step at a time and slowly worked his way through the basement looking for the field. He knew he had found it when he tripped over the field perimeter and fell over onto the tiles, knocking the phone out of his hands. This mission was more hazardous than he had anticipated. Kudos to his phone, though, for all the punishment it had withstood so far today. Hopefully his phone was still functional enough to text.

I'm in the basement and on the field. Power is out and room is black. Looking for the flash drive.

Kenzie replied back immediately. *Power is out here too. Flash drive should be in one of the towers.*

Gordon began searching each tower, looking for the flash drive. Sure enough, he found it in the third tower he checked.

Found it. On my way back.

Wonderful! You're a life saver! Hurry back. Please be safe and stay dry!

Being safe started with finding the field perimeter and *not* tripping over it again on the way out. But it was a bit late to stay dry at this point. He slipped the flash drive into his pocket, found his way back to the door, and went outside. It was important to keep the flash drive as dry as possible and get it back to Corbin ASAP.

Gordon climbed into his car and headed back to Valley Creek High.

* * *

7:42am

"Here ya go," came a voice behind Corbin. In the dark, Corbin couldn't tell who it was until he got closer. It was Gordon, delivering the flash drive as promised.

"Gordon, you rock," said Corbin as he inserted the flash drive into his laptop and restored all of his files and then downloaded the needed code to Columbia. The whole process took less than a minute.

"OK, Emma," said Corbin as he disconnected the cable from Columbia, "it's all yours."

Emma picked up Columbia and went to get in line for inspection.

"Now to deal with Kyle," said Corbin.

Browsing through the files from the flash drive, he found and opened the code for Destroyer that Euggie and Alex had originally swiped for him. Corbin smiled as he made a few creative changes. Luis and Loren looked over Corbin's shoulder as he typed, not really understanding the code he was entering.

While Corbin was typing, Kenzie returned to the pit.

"We're now registered and I turned in our notebook."

"Great," said Corbin, not looking up from his typing as he talked. "Emma is in line waiting for inspection to open."

"So all of our files are restored? And Columbia is working again?" asked Kenzie, looking at a soaked Gordon.

"Yes, I believe so," said Gordon.

Kenzie was so excited to hear this that, without thinking, she stepped over and gave Gordon a big hug.

"Glad to help," said Gordon, a bit shocked, but still pleasantly surprised.

"Hey, Kenzie!" came a voice from the other side of the table. It was Euggie and Cory - and Alex.

Kenzie immediately took a step back from Gordon.

"Hey!" she replied, as unawkwardly as possible. "Nothing like working in the dark, huh?"

"I think it's fun!" said Cory. "It's quite beautiful with all the devices shining. Almost like Christmas lights. And with our late start this morning, we needed the extra time to get set up."

While they talked, Alex was assessing this new guy. First he was with Kenzie under the umbrella. Now he was getting a hug from her. And it hadn't helped matters that this guy had crashed into him earlier outside on the sidewalk.

"Y'all, this is Gordon," said Kenzie, who finally decided to give him an introduction. "Smartest guy in math you'll ever find."

"Oh, we've met," said Alex, as Gordon sheepishly nodded.

"And Gordon just helped us get our files for Corbin to restore on Columbia," continued Kenzie.

"You had to restore your files?" asked Euggie.

"Yes. It's a long story."

"OK, Luis and Loren," said Corbin. "Everything is ready. You know what to do."

Handing him the newly programmed brain, Luis slipped it under his jacket.

"So what exactly did you do to that brain?" asked Kenzie.

"Well, for starters," said Corbin, "it has the updated firmware."

"And?" asked Kenzie.

"Let's just say that Nora is gonna get a big surprise if she tries to freeze or fry a robot."

At that moment, the lights suddenly came back on.

Technicians immediately began making their way to the field control equipment to make sure everything was starting up again and working correctly. At the center of the gym, Coach Warren picked up the mic and announced, "Inspection is now open."

"You all don't have much time," said Corbin.

With that, Luis and Loren left quickly to make their way to Room 108 while Cory and Alex and Euggie went back to the Hobbits' pit to pick up Frodo for inspection.

* * *

8:16am

Watching the two Sonic Boomers leave, Nora decided it would be a good idea to follow them and see what they were up to.

"Robo Em, Nora is now on her way out of the gym."

No answer.

"Robo Em, do you copy?"

Looking around the gym, Jake spotted Emma in line for inspection. She must have her radio turned off. Getting up, Jake ran to the Robonauts' table and right up to Kenzie and Corbin.

"Robo Em has gone dark and we have a development. Nora is leaving the gym," said Jake, standing at attention and pointing toward the door that led into the school.

Kenzie looked over at the door in time to see the Sonic Boomers going out of the gym. Then she watched as Nora looked carefully around the doorway and then slipped through herself. She was following Luis and Loren!

"We have a problem," said Kenzie, running off as quickly as she could towards Room 108.

"Robo Em?" asked Corbin, looking at Gordon.

Gordon shrugged his shoulders.

"But something tells me," replied Gordon, "that Kenzie may need some help."

With that, Gordon left the Robonauts' pit to follow Kenzie into the school.

Chapter 41

8:20am

Luis and Loren entered Room 108 and turned on the light. Kyle's spare brain was sitting right there on the table.

"Let's do this quickly and get out of here," said Loren, "I'd hate for The Best to catch us in here."

Luis went over to the table and carefully made the swap, making sure to put the replacement brain in the exact same position as the one he had just taken. After putting the brain he had swiped into his jacket pocket, they started for the door. Until they saw the doorknob slowly turning. Someone was coming in.

Panicking and not knowing what else to do, Luis and Loren quickly moved behind the door. They held still as the door started to open.

"Hey, Nora," called Kenzie from down the hall. Nora stopped and released the doorknob to Room 108. Turning around, she walked directly towards Kenzie.

"I don't know what you all are up to," said Nora, "but you need to stay out of our business!"

Kenzie paused before replying. "What happened to all the talk about wanting to change? And what happened to being friends?"

Ignoring her questions, Nora demanded, "Here's how things are going to work. You and all the teams you are working with are going to go away and leave me and Kyle alone."

Finally realizing that there was nothing sincere about Nora other than her constant drive to be a pain to everyone else, Kenzie responded firmly. "Or what? What are you gonna do about it? We aren't going to let you all cheat. So you may as well stop trying."

"I'll show you what I'm gonna do about it."

Nora pulled out her phone and hit play.

Immediately, Nora's and Kenzie's voices came out of her phone, talking about Alex.

"You like him a lot, don't you?" said Nora's recorded voice. And, Kenzie's reply, "Yeah, I do, he's really sweet..."

Kenzie stood there frozen, looking into Nora's eyes as this conversation continued to play. Kenzie regretted every word she heard herself saying to Nora on this recording. She would be completely mortified if anyone else heard this.

While all of this was going on, Luis and Loren cracked the door to Room 108 and peeked out.

"I think we can get away if we slip out now," said Luis.

"But where are we gonna go," asked Loren. "We won't be able to get by without Nora seeing us."

"How about the room next door?"

Loren nodded his head in agreement.

As Kenzie stood there in her face-off with Nora, she saw the door to Room 108 open slowly behind her. She watched Luis and Loren from the corner of her eye as they slipped into the robotics lab, undetected by Nora. Once in the robotics lab, Luis texted Kenzie. *Let us know when the coast is clear so we can come out.* Nora was so focused on putting it to Kenzie that she never heard them.

"I have half a mind to post this conversation online!" said Nora as the recording ended. "If you and the teams you are working with don't bug of, the world *will* hear this conversation! Are we clear on this?"

Kenzie didn't get mad easily nor often. But at the moment, she was *furious* at Nora. It was now or never.

Without warning, Kenzie plowed forward as hard as she could, diving directly into Nora and knocking her down. The hit caused Nora's phone to fly out of her hand and land on the floor, skidding a good distance away.

The two girls wrestled on the floor for a few moments before Nora gave Kenzie a hard push, sending her back into the wall. Both girls sat there in the hall for a moment, breathing heavy.

With his head peeked around the corner of the hallway, Gordon had seen and heard all that was going on. And he *had* to do something.

"That's enough!" said Gordon in a loud, authoritative voice as he walked forward and picked up Nora's phone.

Both girls turned their heads to see who was interrupting their quarrel.

"How about we take care of this right now," said Gordon as he made several swipes and presses on Nora's phone.

"And who exactly are you?" asked Nora, standing up.

"Your worst nightmare," said Gordon, his voice cracking a bit.

Nora rolled her eyes for a split second.

"Give it back, you creep!" yelled Nora as she ran toward Gordon in a mad attempt to reclaim her phone.

But there was not enough time. Gordon had already deleted the recording. Permanently.

When Nora reached Gordon, he simply held out the phone for her to take back.

Nora yanked her phone out of Gordon's hand.

"Don't you *ever* mess with Kenzie again!" commanded Gordon.

Nora was so furious she couldn't talk. She simply turned around, marched herself over to Room 108, flung the door open, and went in. The door closed hard behind her with a thud. Inside the room, Nora didn't see any sign of the Sonic Boomers - and it looked like everything Kyle had set up was still in place. She should stay there to guard things. Nora was now more determined than ever to make Kenzie and those pesky Robonauts sorry that they had ever messed with her and The Best.

Out in the hallway, Kenzie wasn't sure what to say, what to do, or what to feel. She was half in shock that she had gone after Nora the way she had. And even more amazed how Gordon had come to her rescue. She and Gordon turned around and began walking back to the gym.

"Thanks," said Kenzie.

"Glad to help," said Gordon, returning to his usual polite self.

"I'm sorry you had to hear that recording. I- "

"Forget about it," said Gordon interrupting her. "Everything is OK now. How about we get you back to the pit and get ready for this tournament?"

Kenzie simply nodded her head yes.

Moments after Kenzie and Gordon had left, Kyle came back through the same hallway with Destroyer. Now that he had successfully completed inspection, it was time to swap the brain so Destroyer would be prepared to freeze and fry robots. When he walked into Room 108 he found Nora sitting there.

"What are you doing here?"

"Guarding everything until you got back. I want to make sure we win *big* today - and that we totally destroy the Robonauts' robot."

Sitting down at the table, it took Kyle about 30 minutes to get the robot brain swapped out. There was no time to test the code because the qualification matches would be starting soon. But there was no need to test. Kyle had downloaded all the needed code to the spare brain the night before and it was ready to go. With the swap made, Kyle packed up everything he had brought to Room 108. Then he and Nora took Destroyer back to the gym for the qualification matches.

Back in the robotics lab...

"Do you think it's safe for us to come out now?" asked Loren.

* * *

8:45am

Inspection was now complete and everyone was starting to gather in the bleachers for the opening ceremonies. Emma and Corbin and Euggie were sitting on the front row and Kenzie was sitting right behind them with Gordon on her left and Alex on her right.

Surrounded by two guys who both liked her quite a bit, it became clear to Kenzie that she had some decisions to make. Alex was sweet, thoughtful, easy to talk to, and shared her love of robotics. On the other hand, Gordon, despite his quirkiness, was kind, polite, and extremely smart. Both would go out of their way to help someone. Yet, in all of this, Kenzie was now realizing that a boyfriend was not what she was looking for right now. Maybe one day. For now, the top priorities in her life were the friends on her robotics team and this incredible adventure they were on together. *This* is where Kenzie wanted to focus her time and efforts. She would break the news to Alex and Gordon later.

"Hey, has anyone seen the Sonic Boomers?" asked Cory who was sitting on the other side of Alex.

Checking her phone for any updates, Kenzie finally saw the text that Luis had sent her.

Yikes! They must still be in the robotics lab! "I'll text them," replied Kenzie.

Luis and Loren returned to the gym right away. Both gave Corbin a fist bump and they climbed past him into the bleachers and took a seat by Cory.

"Good job, you two!" said Cory.

A few minutes later, Nora and Kyle returned to the gym. They looked confident, cocky, and ready for the tournament to get underway.

After the conclusion of the opening ceremonies, the qualification match schedule was distributed. As it turned out, both The Best and the Robonauts were scheduled in the very first match of the day - on opposing alliances. This fateful match would commence precisely at 9:00am.

Chapter 42

As qualification match one started, neither the Robonauts nor The Best were focused on strategy and working with their alliance as much as they should have been. Rather, both were preoccupied with the underlying battle that was about to happen during the match. Nora and Kyle were ready to zap Columbia and any other robot that got in their way. The Robonauts were just waiting to see what Nora and Kyle would do when they figured out that Destroyer was not about to behave as they expected. Euggie had wisely decided to sit this match out and simply watch from the bleachers.

"Remember," whispered Kyle to Nora, "buttons A and B will freeze the opposing robots. And buttons X and Y will fry the opposing robots."

After all safety glasses were in place and both alliances indicated they were ready with a thumbs up, the MC announced, "Autonomous in 3, 2, 1."

Other than the usual preprogrammed routines, nothing out of the ordinary happened. The Best stacked six cubes in their scoring zone, enough for their alliance to take the autonomous bonus points. But in the driver control period, things got interesting.

"OK," whispered Nora to Kyle, "it's time for us to put the Robonauts in their place." Nora put her finger over button A. "Take that," she said under her breath as she pressed it.

But when she pressed button A, none of the other robots on the field froze. Instead, Destroyer suddenly stopped where it was and started spinning in circles, six times around before it stopped.

"What on earth?" screamed Nora.

The Robonauts could barely contain themselves. Indeed, it was hard for Emma to drive Columbia while she was trying so hard not to laugh. Kenzie looked down and Corbin put his hands over his mouth, both in an attempt to hide their huge smiles and continue on as if nothing had happened.

Pressing the button again did no good, as Destroyer started spinning in circles again.

Emma, who was now trying to deliver a stack of ten cubes she had collected, was doing her best to focus. But unable to control herself, she let out a huge laugh. Or was it a snort? Accidentally pressing the joystick a bit too hard, she caused the top three cubes to fall off of her stack and hit the gym floor outside the field perimeter. At least the remaining seven cubes delivered successfully.

Frustrated and not knowing what to do, Nora decided to try pressing Button B. But again, when she did this, no other robots on the field were affected. Just Destroyer. Destroyer stopped again and raised its tray up. Then down. Then up. Then down again. It did this up and down motion ten times quickly, like a dog wagging its tail. At this point, the Robonauts, as well as most everyone in the crowd, were all laughing hysterically.

Nora had not yet tried using the X or Y buttons. Hopefully *this* would work and would fry Columbia's brain. With one minute left in the match, Nora gave her last and best effort. She pressed button X. When she did this, Destroyer started driving straight forward at full speed. It drove right into the Robonauts' scoring zone and crashed through the tower that Emma had stacked. At this point, it was no longer funny and no one was laughing. Immediately, the ref called out, "Illegal entry of the Red Alliance into the Blue Alliance's scoring zone..." But before the ref could get out another word, Destroyer began backing up at full speed. It crashed into the tower in the middle of the field, flinging away every cube that was in its path. Then the robot went forward *again* at full speed, back toward the scoring zone.

“How many times is it going to go back and forth like that?” whispered Kenzie out of the side of her mouth to Corbin.

“Maybe four more times?” replied Corbin, as he shrugged his shoulders.

Kyle reached over, grabbed the controller out of Nora’s hands, and powered it off. When he did this, Destroyer finally stopped moving. The other robots on the field also stopped moving - on their own accord - as each driver’s attention was diverted to what Destroyer had just done. The room was quiet while the clock counted down the last few seconds, ending the match. As the buzzer sounded, everyone stared at the field wondering what had happened. Immediately the ref began conferring with several of his assistants. After a couple of minutes of discussion, they reached a conclusion.

“The Red Alliance illegally entered the Blue Alliance’s scoring zone multiple times.”

After listing a number of additional violations of the Red Alliance, the ref concluded with, “Therefore the Red Alliance is disqualified from this match. Furthermore, the specific offending team is also disqualified from the rest of this tournament and will be referred to the appropriate authorities at REC Foundation to determine their eligibility to participate in future tournaments for the remainder of this season.”

The entire gym remained silent, stunned at what had just happened.

“This robotics stuff is stupid anyway,” screeched Nora, breaking the silence. For the second and last time of the season, she stormed out of the gym. Without any reason to stay behind, Kyle set the controller down, hung his head, and followed her out.

This was the last the Robonauts saw of Nora and Kyle for a long time.

Chapter 43

All the teams of The Resistance made it to the elimination matches at Soaring Heights. The Robonauts, allied with the Sonic Boomers, went all the way to the final match, ultimately winning Tournament Champion. This meant that every team in The Resistance had now qualified for state. Beyond this, the Robonauts received the Excellence award and the Hobbits received the Design award. Team Alpha ended up winning the Design award for middle school. With Team Alpha's focus and level of discipline, they would be a formidable team to contend with in future years as they learned and gained more experience.

But, between the extreme drama leading up to Soaring Heights and all the crazy events of that first qualification match, the rest of this tournament would be relegated to a distant memory for the Robonauts. Kenzie would never ever forget the look on Nora's face as Destroyer was going crazy on the field. Although all the teams of The Resistance did quite well that day, overcoming The Best overshadowed it all. As for that fateful qualification match at Soaring Heights, it became obvious that The Best was clearly and indisputably *not* the best.

Although Kenzie had sincerely hoped that Nora would change, this was not to be. Kenzie would always wonder what it would have been like if Nora had allowed her to be a friend and worked with her instead of against her. Emma was glad she no longer had to worry about Columbia being frozen or fried. In her words, "A driver doesn't need the extra stress." As for Corbin, he had defeated Kyle in all manners of strategic programming - and he was quite proud of himself for doing this.

Now that Nora and Kyle were no longer a threat, Cory and Alex were free to continue developing Frodo for the rest of the season. Cory thought that Kyle and Nora *could* have been an excellent team. With Kyle's programming abilities and Nora's stubborn persistence - two wonderful qualities to have in robotics - perhaps they could have gone far if they had applied themselves to winning with integrity?

Euggie, however, was the biggest casualty in the entire ordeal since he now had no robotics team on which to compete. He continued to hang out with Cory and Alex for the rest of the season, doing his best to learn everything he could about driving, programming, and journaling. At least he was now in a position where he could help and freely learn more about robotics, even if not officially as a team member.

Alex was just glad the whole thing was over. Although he began to enjoy robotics again, his relationship with Kenzie slowed down after Soaring Heights. Especially after Kenzie told him that she only wanted to be friends. They continued to text some, but not as much, and less as the weeks went by.

And then there was Gordon. He had finally made it into Kenzie's good graces. And he didn't seem to mind so much only being friends. After Soaring Heights, Gordon ended up hanging around the Robonauts much more often. He was able to help Corbin with the mathematical aspects of programming and he was able to help Kenzie get through her math homework whenever she needed it. Gordon brought his own quirky personality into the mix. making him a perfect target for Emma's ongoing jokes. At times, it even felt like he was becoming part of the team.

* * *

In the weeks that followed Soaring Heights, changes came that completely rocked the Robonauts' world. The Corona virus that Coach Warren had warned them about had continued to spread and alter the way all public events in society would function. Coach Warren announced the sad news that all further robotics tournaments for the season, including the state championship, had been canceled. There was no robotics news that could have been more disappointing. However, this was the new reality of the world they lived in.

All three of the Robonauts still planned to compete again during their upcoming year at Valley Creek. They had no idea what competitions would look like moving forward, but only that practices like wearing face masks and social distancing were becoming the norm. In truth, no one really knew what to expect. So, the Robonauts did what all good robotics students do when they don't know what to do. They stubbornly decided to move forward no matter what, hit every problem head on, and continue doing their best to learn, be creative, and problem-solve. This started with watching the premiere of the new game for the upcoming robotics season.

Epilogue

Has it started yet? texted Kenzie.

No, we still have about 30 minutes, replied Emma.

Good. This would give Kenzie time to finish up on her project before heading over to Emma's house. Kenzie had printed and framed large pictures of the team from throughout the season including the crazy team picture from Skyscraper. Kenzie was fairly certain that Corbin had not yet seen this. With help from her parents, Emma had mounted a long shelf on the wall in the basement above the practice field that would proudly display these pictures along with all the awards the team had won throughout the year. The half-size field would stay in Emma's basement for the foreseeable future, at least until school reopened again.

Kenzie finished putting all of the pictures into a tote bag and grabbed a six-pack of Dr. Pepper out of the refrigerator, which had somehow become her responsibility to bring to each team get-together. With both of these in hand, she left her house and headed over to Emma's basement. When she arrived, she opened the door and went in to find Emma positioning all of their team's awards on the new shelf.

"Oh, it looks so good!" said Kenzie. Bringing her bag of pictures over, she handed them to Emma one by one. Emma positioned each of these on the shelf in between the team awards.

Stepping back to admire their work, Emma had a tear in her eye. "You know, we went through a lot together this past season." She looked at a picture of the whole team receiving the Excellence award at Skyscraper and the crazy picture where Corbin was positively identified as the "Team Geek". Centered and prominent in the center of the shelf was a picture of the entire team with Columbia taken at Soaring Heights, their school's very first robotics competition.

Many of the team's memories, however, were not captured in pictures. Kenzie thought back to all the drama with Nora and Kyle. She thought back to the special evening she had spent with Alex the evening after Skyscraper. And, who could forget all the hours of work that had been put in at Pietro's, the school, and Emma's basement.

It would be terribly hard on the Robonauts when the time came to take apart Columbia. Even though Columbia was only a compilation of metal, plastic, and electronics, it had somehow become more to them. Indeed this robot had been one of their primary focuses for the entire season. But in the end, they would need the parts from Columbia to put together a new and different robot designed specifically for next year's game.

As the girls were standing there reminiscing, someone knocked on the basement door.

"Come on in," yelled Emma.

The girls watched as Corbin and Gordon came through the door. Immediately Corbin set down the backpack he always carried, pulled out his laptop, and sat down with it on the field. "How much time do we have?" he asked.

"About 10 more minutes," said Emma.

While Corbin was getting set up, Gordon had joined the girls in looking at all the awards and pictures now displayed on their new shelf.

"So, you all are going to have to explain to me how this *reveal* works," said Gordon. Even though Gordon felt much like part of the team, he really had very little experience with robotics and all of this was still new to him.

"Every year," said Kenzie, "a new robotics game is announced by VEX Robotics and REC Foundation. This past year was Tower Takeover and we're about to find out what next year's game is."

"Normally they reveal the new game at the end of Worlds each year," added Emma, "and this year is no exception. The only difference is that, since the World Championship was canceled for this season, they are having a virtual celebration instead, with the game reveal to follow."

"I hope there will be a Worlds tournament next season," said Kenzie, "because I *really, really* want to go."

"But we can only go if we are able to qualify," said Corbin from the floor.

The reveal was a big moment for the Robonauts because they were about to find out what types of tasks would be occupying their minds and their time for the next year. This past year had been all about cubes and how to collect them and stack them or put them in towers. But what would they have to figure out how to do for this coming year? The big game reveal was truly an exciting time for *everyone* involved in robotics.

Corbin now had the VEX YouTube channel playing and the live broadcast of the reveal would start in a few minutes. Standing up, he walked over to the shelf and looked at each picture and award, one by one.

"Hey! What's this? Team Geek? Really?"

Kenzie and Emma burst out laughing. "You know we love ya," said Emma, "but we couldn't resist."

Corbin turned around quickly when he heard the words coming from his laptop, "And now, the reveal of the new game..."

"Here it is!" said Corbin excitedly.

The Robonauts all gathered around the laptop, listening intently. The new game was called "Change Up." It would be played on the same size field, but with different field elements and game elements. Rather than the towers and scoring zones they had been accustomed to working with, they would now be working with nine round goals that would be positioned around the field. Four of these goals would sit in the corners, four would sit on the sides of the field, and one would sit smack dab in the center. Combined, all nine would form the shape of a tic-tac-toe board. And there would be numerous red and blue dimpled balls on the field. The task of each team would generally be to build a robot that picks up the balls and puts them in the goals in various arrangements to score points. As always with VEX Robotics, tons of strategy would be involved and there would be numerous details to work through if the Robonauts were going to put together a top-quality, competitive robot.

As the presentation for the reveal ended, ideas for how a robot might pick up the balls were already swirling around through Emma's head. Kenzie was already sketching out ideas for various mechanisms for use in this game. Corbin was busy pulling up the game manual on his laptop. Gordon was watching and learning and trying to absorb all that was going on here.

"So are you all ready?" asked Kenzie.

"Not yet," said Emma, "but I will be any minute."

At that moment, there was a knock at the basement door. Running over to answer it, Emma opened it and exchanged a handful of money for a large Pietro's pizza and a basket of hot wings.

"Thank you," said Emma, closing the door. Then she turned around. *"Now,* we are ready!"

"Ready for what?" asked Gordon.

All three Robonauts looked at Gordon with a look that said, *Duh?*

"Planning the robot for next year, of course!" said Kenzie, as she broke out the Dr. Pepper.

As they all sat down to eat and start planning, Corbin's phone lit up. It was a text from Coach Warren. "Guess what you all! Coach says he just placed the order for the Change Up game pieces and field elements. They should arrive in a few weeks."

Gordon picked up the basket of hot wings and took a whiff of them. "Are these very spicy?"

"Nah, not really. You ought to try one," said Emma with a quick wink at Corbin.

Gordon set them down. He might have to try one in a bit.

The conversation then turned to topics such as how to pick up a ball. What type of intakes should be used? What speed would the motors need to be? Once the robot picked up the balls, how would the robot raise them high enough to get them to the top of the goals? Then, once the ball was high enough, how would the robot deliver the ball into the goal? The questions and problems that needed to be solved were endless. But the Robonauts were up for the task. They listed out parts, drew up diagrams, talked about how they might use sensors, and even got some good suggestions from Gordon on how to incorporate geometry to get some measurements that would be needed.

Maybe math wasn't so bad after all? thought Kenzie.

Suddenly Corbin's laptop starting ringing with an incoming video call. It was Cory and Alex and Euggie in one window - and the Sonic Boomers in another window.

"I thought we ought to have one final meeting of The Resistance," said Cory. "So what do you all think about Change Up?"

"I think it's gonna be fun," said Luis.

"But it's definitely going to be a shift in our thinking to work with balls instead of cubes," said Emma. "Cubes sit there. But balls roll. How are we supposed to pick up an object that won't even stay still?"

"Oh, that's easy," said Euggie. "We can stop the ball from rolling by-"

"Noooo," interrupted Alex. "Everyone else can figure that out on their own."

Alex looked in the camera and smiled. "Confidential team info."

"Did I hear you say, *we?"* Kenzie asked Euggie.

"Yes, you did," replied Euggie.

"Euggie's going to be joining our team this year as our official driver," announced Cory. "This will let me spend more time journaling - and we'll work together to build the robot."

"And, I'm going to focus on programming," said Alex. "But don't worry, I won't try to freeze or fry any robots."

"If you do, I'll be coming after you," said Corbin.

"It looks like we aren't the only team with a new member," said Cory.

All eyes looked at Gordon.

"Um, well I..." Gordon stammered.

"Oh yeah, Gordon is now on our team," said Kenzie. And this spontaneous comment made it official. The Robonauts now had four members.

"You see this?" said Kenzie, holding up the Dr. Peppers in front of Gordon. "These are considered essential for every team meeting. And I now pass this critical responsibility on to you."

Corbin nodded his head in approval.

Gordon was in for a fun year.

After chatting awhile longer, the final meeting of The Resistance came to an end. Although the Robonauts knew the coming year would have many challenges with many ups and downs, they were ready to plan their attack and rise up to meet those challenges. With a sense of excitement and anticipation, they worked hard for the next 8 hours. It was the first of many late nights to come as they embarked on the grand adventure of the new VEX Robotics season.

Further Information About VEX Robotics

If you are already involved with VEX Robotics, I hope you have enjoyed reliving many of the aspects of the life of a robotics team through this story. If you would like to learn more about VEX Robotics, you can find tons of information at the following websites. Although this story is centered on VEX Robotics Competition (VRC), the program that encompasses grades 6-12, other competitive robotics programs are available for participants as young as Pre-K and as old as college.

roboticseducation.org

This is the official site of REC Foundation (Robotics Education & Competition Foundation) and the first place I would recommend that you go to learn more about competitive robotics. According to this web site, *"The REC Foundation's VEX family of robotics programs for elementary school through college students includes 24,000 teams in over 60 countries. These engaging programs offer a direct response to workforce and industry needs with participants more likely to consider studying science, technology, engineering, and math (STEM) beyond high school."*

vexrobotics.com

This is the official site of VEX Robotics where you can go to purchase robot kits and parts that are competition-legal.

robotevents.com

This site is the central hub where all scheduled tournaments and other events are posted. Coaches and teams use this site to register teams for events and event planners use this site to communicate with participants. Ongoing scores and statistics of registered teams are posted here and students often use this site to view their various rankings as well as track the progress of other teams.

vexforum.com

This site is a community of robotics students where questions, problems, etc. may be posted. Robotics students often spend regular time browsing this forum looking for ideas and exploring what others are doing with robotics. This site is also one of the best places to get assistance and technical support.

Made in the USA
Monee, IL
04 November 2026

67387875R00115